Quantum AI

The Future of Intelligence

By

Aiden Cooper

Quantum AI

The Future of Intelligence

Table of Contents

Introduction

In a world increasingly driven by rapid technological advancements, quantum computing and artificial intelligence stand out as both pioneers and revolutionaries. These potent technologies promise not only to redefine the limits of computation and intelligence but also to revolutionize industries, reshape societies, and tackle some of the most pressing challenges we face today. This book aims to serve as a compass, guiding you through the intricate landscape of quantum computing and AI, unveiling their fundamentals and exploring their far-reaching potential.

The fabric of today's technological ecosystem is woven with possibilities that were once confined to the realms of science fiction. Quantum computing, with its fundamental reliance on the bizarre principles of quantum mechanics, defies traditional logic and expands the boundaries of what's computationally feasible. Simultaneously, artificial intelligence, with its ever-evolving capabilities, promises to augment human decision-making, automate complex processes, and even enhance creative endeavors.

As we embark on this journey, it's important to appreciate the diverse avenues through which quantum computing and AI intersect. Each chapter in this book navigates a specific domain where these technologies converge, offering insights and fostering understanding. We'll explore their theoretical underpinnings, practical applications, and the societal implications of their integration. By understanding how and why these technologies function, we can better anticipate the future they herald.

Over the next few pages, we'll discuss the advent of quantum computing and AI, focusing on how these systems can perform feats that surpass the capabilities of classical approaches. Today's supercomputers, while extraordinary, operate on principles that quantum computers promise to transcend. The future implications of quantum-enhanced capabilities are vast, spanning everything from cryptography to climate modeling and beyond. Understanding these potential shifts is crucial for anyone aiming to harness—or even simply comprehend—what the next few decades may bring.

Yet, with great power comes the responsibility to wield it wisely. This introduction sets the stage for discussing not only the technical aspirations of these technologies but also the ethical considerations and societal shifts they may precipitate. In delving into quantum AI, we must remain mindful of its double-edged nature, capable of both creating solutions and presenting challenges. This dynamic landscape is filled with opportunities for innovation but also requires vigilance and foresight to navigate responsibly.

The narrative of technological evolution is complex and multifaceted, one that is deeply intertwined with human curiosity and the relentless drive to explore the unknown. Each leap forward represents a culmination of interdisciplinary collaboration, scientific breakthroughs, and the collective ambition to improve the human condition. As this journey unfolds, we find ourselves on the brink of a new era—one where quantum computing and AI don't just complement each other but converge and amplify their respective strengths, propelling us towards new horizons.

This book endeavors to demystify these futuristic tools. Our goal is not only to inform but also to inspire—a purpose that reflects both an intellectual curiosity and a hopeful vision of what these advancements can achieve. As we peel back the layers of these intricate topics, we aim to cultivate a sense of awe for the progress made thus far and the potential yet to be realized. Join us as we seek a deeper understanding, not just of the technology itself, but of its transformative potential to redefine what it means to live in a world informed by quantum and AI capabilities.

With concepts like quantum superposition, entanglement, and neural networks, we'll unravel the threads that connect microscopic phenomena with macro-scale applications. We'll pose open-ended questions that challenge the perceived limits of computing and intelligence, fostering a dialogue that extends beyond the pages of this book. As you engage with the material, consider not only how these technologies might solve current problems but also how they might alter the nature of those problems altogether.

In preparing this book, we've drawn upon recent research and insights from leading experts across the globe, ensuring a comprehensive and up-to-date exploration. The field is actively evolving, with new discoveries and innovations emerging regularly. This introduction lays the groundwork for a dive into a world where the conventional and the extraordinary blend seamlessly, offering you a glimpse into what lies ahead.

Ultimately, the future of quantum computing and artificial intelligence is not merely a matter of technological progress but one of visionary thinking and global collaboration. By understanding their foundational principles, recognizing their potential applications, and considering their broader implications, we can engage with these technologies more effectively. This foundational perspective will serve as a keystone for the chapters that follow, each offering a deeper exploration into the burgeoning fields that quantum computing and AI are beginning to define.

Chapter 1:
Understanding Quantum Computing

Diving into quantum computing means venturing into a realm where the very fabric of computational reality is being rewritten. Building upon the foundations of quantum theory, this revolutionary field harnesses the strange yet powerful principles of quantum mechanics, promising to outpace classical computing by magnitudes unimagined by even the most vivid tech dreamers. At the heart of this leap are qubits, the quantum bits that exist in a fluid state of superposition, defying traditional binary constraints and enabling the mind-bending phenomenon of entanglement. Together, these elements open doorways to computation with potential applications spanning cryptography, optimization, and data analysis. As we decipher these quantum codes, we're not simply gaining a new toolkit; we're poised on the brink of a technological metamorphosis that could fundamentally transform industries and tackle complex global challenges. In this chapter, we start the journey of understanding how these quantum elements power a new era of computational capability.

The Basics of Quantum Theory

In exploring the foundational structure of quantum computing, we must first delve into the basics of quantum theory, a framework that redefines our understanding of the universe at the smallest scales. Quantum theory introduces a new realm where particles can exist simultaneously in multiple states, defying conventional logic and opening doors to unprecedented computational possibilities. This duality and interconnectedness fundamentally contrast the deterministic nature of classical physics, leading to the emergence of phenomena like superposition and entanglement. These concepts, while initially counterintuitive, underpin the transformative potential of quantum computing, promising to solve complex problems with an efficiency that classical computers can't achieve. As we grasp these quantum principles, we unlock a world rich with opportunity and mystery, poised to challenge our perception of reality and reshape the technological landscape.

Key Principles of Quantum Mechanics are the foundation on which the enigmatic and transformative world of quantum computing is built. At the heart of The Basics of Quantum Theory, these principles challenge the way we understand reality itself. For those immersed in classical physics, the leap to quantum mechanics is profound. Where Newtonian mechanics offered predictability, quantum mechanics presents probabilities, transforming certainty into a realm of possibilities.

The most central principle is quantization. In the quantum realm, energy is not continuous but comes in discrete packets known as quanta. This principle was introduced by Max Planck in the early 20th century to explain blackbody radiation, setting the stage for a new era in physics. It implies that at the microscopic level, energy transitions occur in jumps rather than smooth changes, a departure from classical expectations.

Another foundational concept is wave-particle duality. First brought to light by Louis de Broglie, it posits that every particle exhibits both wave and particle properties. This dual nature is famously illustrated in the double-slit experiment, which demonstrates that solitary particles like electrons create interference patterns—a hallmark of wave behavior—under certain conditions. This duality challenges our classical perception of particles as solid and individuated.

Perhaps the most perplexing of all is the principle of superposition. Superposition allows quantum systems to exist in multiple states at once until observed or measured. This means a quantum bit, or qubit, can be both 0 and 1 simultaneously, enabling quantum computers to process vast amounts of data concurrently. Schrödinger's cat, a thought experiment, elegantly illustrates this phenomenon by placing a hypothetical cat in a state of being both alive and dead until its condition is observed.

Entanglement is another groundbreaking principle, one that Albert Einstein famously dubbed "spooky action at a distance." When two particles become entangled, the state of one instantly influences the state of the other, regardless of the distance separating them. This interconnectedness defies classical intuitions about locality and demonstrates a profound interconnectedness in quantum systems. Entanglement is key to quantum computing, enabling qubits to perform complex calculations in tandem.

Adding to the complexity is the uncertainty principle, articulated by Werner Heisenberg. It posits that certain pairs of physical properties, such as position and momentum, cannot be simultaneously known to arbitrary precision. The more precisely one property is measured, the less precisely the other can be controlled or predicted. This inherent uncertainty is not due to limitations of measurement but is a fundamental property of quantum systems. It challenges deterministic views and indicates that at a fundamental level, reality is fuzzier than once thought.

The mathematical framework underpinning quantum mechanics is dense but beautifully intricate. It relies heavily on linear algebra and complex probabilities. Operators, wave functions, and matrices form the toolkit for predicting the behavior of quantum systems. These mathematical constructs drive the algorithms of quantum computing, allowing these machines to tackle problems once thought intractable.

Despite their abstract nature, these principles have concrete implications. Quantum tunneling, for instance, relies on the probabilistic nature of particles allowing them to pass through barriers. This principle not only underlies the mechanics of semiconductors and diodes but also opens paths toward the development of quantum tunneling devices and more efficient forms of energy production.

The debates and discussions surrounding the interpretation of quantum mechanics continue to inspire. From the Copenhagen interpretation, which emphasizes the role of the observer in determining the state of a quantum system, to the many-worlds theory that suggests each quantum event spawns new, parallel realities, the philosophical implications of quantum mechanics are as profound as their scientific ones. Each interpretation offers unique insights into the fabric of reality while fueling new lines of scientific inquiry.

However, to truly grasp the implications of these principles, one must embrace their counterintuitive nature. They demand a shift in thinking, pushing beyond the comfort of the macroscopic world into the unpredictable dance of particles and waves. As the foundations of quantum computing, they pave the way for innovations that promise to not just solve computational problems at an unprecedented scale but also redefine industries and human potential.

Understanding these principles is a journey through a world where the rules are still being written and discoveries await just beyond the horizon. In the context of quantum computing, they represent more than mere rules; they are the keys to unlocking a future where technology transcends its own boundaries, offering solutions to the most daunting challenges of our time. Scientists and technologists stand at the cusp of this frontier, armed with an understanding of quantum mechanics and the determination to see where this revolution in thought and capability will lead.

Quantum Bits: Qubits Unveiled

As we dive deeper into the world of quantum computing, one encounters the fascinating concept of quantum bits, or qubits. While classical computing relies on bits as the fundamental unit of information, which exist in one of two definite states—0 or 1—qubits operate under the strange yet profound principles of quantum mechanics. This makes them a cornerstone of quantum computing, promising to revolutionize how we process information.

Unlike classical bits, qubits harness the principles of superposition and entanglement, which endow them with capabilities far beyond their classical cousins. Superposition allows qubits to exist in multiple states simultaneously, not just 0 or 1 but both at the same time. This characteristic exponentially increases the computational power available when multiple qubits interact. Consider it akin to a coin spinning rapidly through the air, embodying both heads and tails at once, representing a multitude of potential outcomes rather than a binary choice.

Moreover, qubits can become entangled, a phenomenon described by Einstein as "spooky action at a distance." Entanglement connects qubits in such a way that the state of one instantly influences the state of another, regardless of the distance separating them. This connection is fundamental to the impressive parallel processing potential inherent in quantum computers, allowing them to solve problems in entirely novel ways.

Building qubits presents a unique set of challenges that stem from their sensitivity to external influences. Ensuring quantum coherence, which is vital for the proper functioning of qubits, demands a nearly perfect isolation from external environmental factors. Even the slightest interference can lead to decoherence, disrupting the qubit states and corrupting the computations relying on them.

However, the effort dedicated to overcoming these obstacles is matched by the potential rewards. Quantum supremacy—a term used to describe the point at which quantum computers surpass the abilities of the most advanced classical computers—relies heavily on the successful manipulation of qubits. Achieving this milestone could unlock the power to tackle complex problems that remain unsolvable by today's traditional machines.

The architecture of qubits is as varied as the theoretical landscapes of quantum mechanics itself. Qubits can be realized through various physical systems, such as trapped ions, quantum dots, and superconducting circuits. Each implementation exhibits distinct advantages and disadvantages concerning coherence times, scalability, and error rates. Consider trapped ions, which offer high coherence times and accuracy, yet often fall short on scalability compared to superconducting qubits, which have emerged as a prevalent option due to the balance they strike between coherence and integration into existing technological frameworks.

Superconducting qubits, for example, have become a focal point of industry giants such as IBM and Google, who are pouring resources into scaling up the number of qubits in their quantum processors. By continually pushing the boundaries of qubit integration, they aim to unlock the complete potential for error-corrected quantum computation. Error correction is crucial—it transforms susceptible qubits into reliable building blocks, enabling the construction of robust quantum systems capable of sustaining long algorithms and complex calculations.

Yet, the journey of taming qubits is far from over. The scientific community faces a dense frontier, where theoretical advancements must go hand-in-hand with experimental breakthroughs. Every new discovery brings us closer to a future where qubits are as ubiquitous as transistors are today. The practical implications of deploying functional quantum processors are vast, offering avenues to advance numerous fields ranging from cryptography and material science to optimization problems and beyond.

In the context of cryptography, quantum bits could upend current encryption protocols thought to be impenetrable. Shor's algorithm, a quantum counterpart of conventional factorization methods, implies that once quantum computers scale adequately, they could decode RSA encryption with unprecedented speed, necessitating sophisticated quantum-resistant algorithms to safeguard digital communications.

Beyond cryptography, the potential of qubits extends to simulating molecular structures and chemical reactions—a task where classical computers hit their limits due to the sheer complexity and number of variables involved. Firms and academic institutions eagerly anticipate using such simulations to revolutionize drug development and material innovations, unraveling secrets that have puzzled scientists for decades.

This revealing journey with qubits doesn't tread a solitary path; it crosses multidisciplinary boundaries. Collaborations across physics, computer science, and engineering are intrinsic to furthering our understanding and harnessing the power of qubits. Interdisciplinary teams combine expertise to refine quantum algorithms, optimize error correction, and push forward our architectural understanding of scalable quantum systems.

While the road ahead is laced with challenges, it is this pursuit that feeds the relentless curiosity and sheer ingenuity driving the field of quantum computing. The promise of qubits inspires a paradigm shift in computational possibilities, painting a future where the fabric of modern industries, from data processing to AI, is woven with the threads of quantum communication and calculation. With each stride forward, qubits not only unlock novel solutions but also invite us to rethink conventional boundaries, igniting a new era in technology.

Superposition and Entanglement: A Closer Look

Quantum computing stands apart from classical computing through its two cornerstone phenomena: superposition and entanglement. These concepts, in essence, are the lifeblood of quantum technologies. Although their implications might initially seem abstract, they hold the potential to revolutionize how we compute and process information, making them crucial to understanding the transformative capacity of quantum computing.

At the heart of superposition lies a radical departure from classical binary logic. In classical computing, bits exist in a state of 0 or 1. Quantum bits, or qubits, on the other hand, exist in a superposition of 0 and 1. This means that until we measure a qubit, it effectively holds both potential values simultaneously. This duality is not just a theoretical construct; it lays the groundwork for performing calculations at speeds unattainable by classical computers. Imagine asking not just a single question of the universe at a time, but exploring a myriad of possibilities simultaneously. It's this ability to represent and manipulate multiple states at once that imbues quantum computers with their unrivaled power.

The underpinnings of quantum superposition challenge our classical intuition, driving us to understand concepts in ways we never needed to with classical physics. Schrödinger's famous thought experiment—the cat that is simultaneously alive and dead—serves as an allegory to illustrate the counterintuitive nature of quantum mechanics. When we scale this concept up to quantum computing, it becomes evident how superposition allows a profound leap in computational capabilities. An exponential increase in states under examination facilitates the solution of complex problems, such as factoring large numbers or simulating molecular interactions. This capacity heralds a new era of problem-solving.

Where superposition allows qubits to explore numerous states, entanglement imposes a unique type of order or correlation. When two or more qubits become entangled, the state of one qubit instantaneously influences the state of another, no matter the distance between them. This non-local property defies classical understanding and has profound implications for the way information is transferred and processed. The influence of one entangled particle on another is instantaneous, briskly surpassing the speed limits set by the speed of light, challenging the very foundations of relativity and communication.

Entanglement delivers unprecedented opportunities for quantum communication and cryptographic protocols. Through a process known as quantum teleportation, entangled states enable the transfer of information in ways that are fundamentally secure against eavesdropping. It's akin to having a key that, once used to unlock the box, is no longer usable by anyone else. This transforms the domains of secure information transfer and cyber defense, promising safety and confidentiality on levels that classical systems struggle to achieve.

Bringing these phenomena together in the form of a quantum computer offers us new lenses through which to tackle problems such as optimization in complex systems, machine learning patterns that classical computers can't discern, and the simulation of quantum systems themselves, which has lengthy ramifications in fields as broad as pharmaceuticals, materials science, and beyond. The computational world, brimming with new possibilities, awaits the clarity that only these quantum phenomena can provide.

The challenges that accompany harnessing superposition and entanglement are not trivial. To sustain a qubit's state of superposition without decoherence, or the unwanted collapse into randomness, requires environmental isolation and innovative error correction methods. Entanglement, too, demands precision manipulation and control over quantum systems that operate at the limits of current technological capability. This is why we are still in the nascent stages of fully realized quantum computing and why interdisciplinary efforts are crucial in overcoming these technical barriers.

Nonetheless, the spirit of innovation and discovery propels us forward, driven by the potential to revolutionize fields from cryptography to artificial intelligence. As we push the boundaries of what's possible, considering superposition and entanglement isn't merely an academic exercise; it's about preparing for a future where quantum computing doesn't just complement classical computing but unlocks new dimensions of informational potential. Our quest for understanding these phenomena isn't merely to grasp their theoretical allure, but to wield them as tools for change and progress.

The integration of superposition and entanglement into computation challenges not only our technological boundaries but also our philosophical perspectives on reality and information. These concepts urge us to reconsider the foundational principles of how we interpret the universe and the role of information within it. As we strive to extract maximum utility from quantum mechanics, we edge closer to unveiling new capabilities that have long been shackled by the limitations of classical thinking.

In essence, understanding superposition and entanglement offers us a portal into the future—a future where the enigmatic and the tangible merge, paving the way for infinite possibilities grounded in the peculiar and captivating logics of quantum mechanics. As these principles become more intricately woven into the fabric of information processing, they promise not just to redefine computation but to fundamentally transform the paradigms of technology and intelligence, propelling humanity into a new epoch of innovation and discovery.

Applications in Modern Computing

Quantum computing, once a mere theoretical playground for physicists, is now carving niches in various domains of modern computing. Its potential to revolutionize industries stems from its ability to process complex problems at speeds unattainable by classical computers. Traditional binary computers utilize bits, which exist in a state of 0 or 1. Quantum computers, on the other hand, exploit the enigmatic properties of qubits, allowing them to exist in multiple states simultaneously, thanks to superposition. Such properties provide a profound advantage in computational power and efficiency.

Simulating molecules and materials, one of the most computationally intense tasks, showcases how quantum computing is reshaping the frontiers of modern computing. Quantum computers have the potential to solve complex quantum chemistry and physics problems that are currently intractable for classical computers. This capability can accelerate the discovery of new materials and drugs, bringing breakthroughs in everything from battery technology to pharmaceuticals. For instance, simulating interactions in a caffeine molecule may seem trivial but requires an astronomical amount of computing power. Quantum computing's spin on the issue could, metaphorically, turn this computational mountain into a molehill.

Optimization problems are another promising application of quantum computing. These are prevalent in logistics, financial modeling, and machine learning. Classical computers must often rely on heuristics to solve them, potentially missing the optimal solution or taking significant time to compute. Quantum computers, leveraging quantum entanglement and superposition, can analyze every potential solution simultaneously, identifying the most efficient path forward quicker than traditional systems. For example, airlines could use quantum computing to optimize flight routes and schedules, reducing fuel consumption and improving travel efficiency. The resulting impact isn't just ecological but economic, offering significant savings.

There's a robust intersection of quantum computing with artificial intelligence (AI) that promises a future where AI systems operate with unprecedented efficiency and insight. Quantum Machine Learning (QML) is an emerging field focusing on employing quantum computing power to enhance machine learning algorithms and models. Current AI applications are limited by computational power; with quantum computing, algorithms could become vastly more adept at recognizing patterns, natural language processing, and decision-making tasks. Imagine AI systems capable of processing data in real-time at levels that surpass human cognitive capabilities, leading to breakthroughs in fields like autonomous vehicles and climate modeling.

Moreover, quantum computing's inherent advantages in handling complex simulations and optimizations make it an invaluable tool in sectors like telecommunications and network architectures. As data traffic increases and network structures become more intricate, quantum computing can manage data routing and loading tasks with higher precision and speed, ensuring seamless communication pathways. Not only does quantum computing offer enhanced performance capabilities, but it also brings transformative potential to cybersecurity frameworks. Quantum cryptography, with its promise of virtually unbreakable encryption, could redefine how secure communications are conducted globally.

However, amid such transformative potential lies an unending string of technical and theoretical challenges. Error rates and quantum decoherence are significant obstacles, but the progress in quantum error correction techniques is encouraging. As researchers refine these technologies, quantum computing's integration across sectors could pioneer new digital paradigms, challenging our fundamental notions of computational limits.

In conclusion, quantum computing's applications in modern computing represent not just an evolutionary step but a catalyst for profound change. Whether optimizing powerful AI systems or securing our communications through quantum encryption, the future beckons an era where quantum mechanics amplifies our technological horizons. The full realization of these applications in everyday computing is poised not just to enhance existing technologies but to pioneer entirely new domains of human cognition and connectivity.

Chapter 2: Foundations of Artificial Intelligence

Building upon the groundbreaking paradigms that revolutionized modern computing, the foundations of artificial intelligence (AI) emerge as a transformative force within the technological landscape. As we delve into its historical evolution, we discover how AI's roots stretch back to the ambitious dreams of domain pioneers who sought machines that could solve complex problems with human-like intuition. Through a tapestry of innovations in machine learning and neural networks, AI systems have graduated from visionary concepts to powerful entities capable of tasks deemed previously insurmountable. By juxtaposing AI and human intelligence, we uncover not just the complement and contrast between the two but also the potential for AI to augment human capabilities beyond traditional boundaries. This exploration sets the stage for understanding how AI will interlace with emerging disciplines, fostering avenues for unimagined advancements. Such a foundation not only demystifies the engines driving AI but also inspires curiosity about the unprecedented futures they herald.

The Evolution of AI Technology

Artificial Intelligence has evolved from a fledgling curiosity into a formidable force that's transforming industries, society, and the very fabric of our technological landscape. At its core, AI is rooted in the simulation of human intelligence processes by machines, especially computer systems. These processes include learning, reasoning, problem-solving, perception, and language understanding. The journey of AI began over seven decades ago and has seen incredible transformations, driven by computer advancements, theoretical breakthroughs, and the relentless pursuit of innovation.

The origins of AI can be traced back to the mid-20th century. In 1956, a conference at Dartmouth College laid down the idea that human intelligence could be precisely described so that a machine could simulate it. This marked the beginning of AI as an academic field. Early efforts in AI were limited by the hardware and the lack of large datasets. Programs were created to solve algebra problems, prove theorems, and even play games, yet these were brittle and lacked robustness outside their designed framework.

Despite these limitations, by the 1970s and 1980s, AI research had produced numerous breakthroughs. Expert systems, which tried to emulate the decision-making ability of a human expert, became popular. These systems were utilized in business and medical applications, contributing significantly to process improvements. However, these early systems lacked the ability to learn and adapt without significant human intervention. The high cost and low performance led to the "AI Winter," a period when funding and interest in AI waned considerably.

AI experienced a significant resurgence in the 1990s with the development of machine learning and neural networks. This era embraced a statistical approach, focusing on creating algorithms that could process vast amounts of data and learn from it. Unlike expert systems, machine learning offered flexibility and adaptability. Algorithms could be trained on datasets to achieve high accuracy without manual rule programming. This shift laid the foundation for today's AI technologies, which emphasize learning and adaptation.

The explosion of data in the digital age dramatically accelerated AI's evolution. The internet, social media, and the proliferation of smart devices created an unprecedented amount of data. Machine learning, specifically deep learning, with its ability to extract complex patterns from large datasets, became a pivotal technique in AI's toolkit. Convolutional neural networks and recurrent neural networks advanced the fields of image recognition and natural language processing, revolutionizing sectors ranging from healthcare to finance.

The progress in hardware, particularly Graphics Processing Units (GPUs), massively parallel computing, and more recently, Tensor Processing Units (TPUs), enabled the training of complex models that were once computationally prohibitive. These advances in hardware are as crucial as software breakthroughs, providing the horsepower needed to train neural networks efficiently. The synergy between software algorithms and hardware capabilities continues to push AI to new heights, managing tasks once considered purely human.

Today, AI is ubiquitous, embedded in various facets of life—from autonomous systems and predictive analytics to personal digital assistants and recommendation engines. The improvement in AI technology is not just a result of algorithmic advancements but also the integration of interdisciplinary innovations, including cognitive science, neuroscience, and psychology, to understand and mimic human thought processes better. This holistic approach not only enhances AI's capabilities but also makes its applications more intuitive and human-like.

The landscape of AI is continuously evolving, driven by emerging technologies and research. Reinforcement learning, a method where an agent learns to make decisions by taking actions in an environment to achieve some notion of cumulative reward, is gaining traction. This method has been instrumental in developing AI that can play complex games and manage real-world uncertainties. It opens new frontiers in dynamic and adaptive system development, a critical component for applications like robotic automation and smart grid management.

AI's evolution also involves ethical considerations and societal impacts. As AI systems assume more roles traditionally occupied by humans, questions about accountability, privacy, and employment arise. This necessitates ongoing dialogue on ethical AI development and deployment, ensuring that technological advancements benefit humanity as a whole. The future of AI will not just be about technological progress but will also involve navigating these profound ethical and humane considerations.

Looking forward, AI technology holds the promise of breakthroughs that could redefine what machines can achieve. Advances in explainable AI aim to demystify the decision-making processes of complex models, making them more transparent and accountable. This is crucial in high-stakes areas like medicine and finance, where understanding AI decisions is not just advantageous but imperative. Collaborative AI, which emphasizes cooperation between humans and machines, represents another exciting frontier, enabling systems that can augment human capabilities, enhancing productivity and creativity across domains.

At the cutting edge of AI research, quantum computing looms large, promising a paradigm shift in computational capabilities. Quantum computers, leveraging superposition and entanglement, could solve problems intractable for classical computers, offering new horizons for AI development. These quantum-enhanced AI systems could revolutionize fields such as drug discovery, materials science, and climate modeling, making possible what was previously unattainable.

AI's story is one of relentless innovation and adaptation. As it continues to evolve, each advancement brings new opportunities and challenges. By learning from the past and integrating insights from diverse fields, AI's future will be shaped not just by technological prowess but by our collective ability to harness its potential responsibly and ethically. In the coming years, the evolution of AI technology will be marked by collaborations across disciplines, integrating innovations from quantum computing and other emerging fields to create solutions that are smarter, faster, and more impactful than ever before.

Machine Learning and Neural Networks

Machine learning and neural networks form the backbone of modern artificial intelligence, bringing scientific imagination closer to technological reality. These technologies have not only transformed industries but also reshaped how we perceive intelligence itself. Understanding their foundational principles and methodologies is crucial to grasping the larger tapestry of AI advancements. We are standing on the shoulders of mathematical giants whose theories and equations made it possible to transform vast amounts of data into actionable insights.

Machine learning is, at its core, a subset of artificial intelligence that focuses on building systems that can learn from data, identify patterns, and make decisions with minimal human intervention. Unlike traditional programming, where explicit instructions are given to the computer, machine learning involves creating algorithms that allow the machine to create its own model of understanding based on input data. From recognizing patterns in streaming services to recommending the next purchase, machine learning is all around us, quietly working behind the scenes.

There are several types of machine learning, tailored to different kinds of tasks. Supervised learning involves training an algorithm on a labeled dataset, meaning the output is already known. This model of learning is often used in classification and regression tasks. In contrast, unsupervised learning deals with unlabeled data, and the system must infer the natural structure present within a set of data points. Finally, reinforcement learning differs from both because it focuses on making a sequence of decisions while optimizing for a particular goal.

Neural networks, inspired by the human brain, are pivotal in the domain of machine learning. They consist of interconnected nodes, or "neurons," which work together to process complex data inputs. These neurons work by taking an input, processing it, and passing on the information to the next layer of nodes. The strength of neural networks lies in their ability to model intricate relationships and recognize patterns that are largely invisible in traditional data analysis methods. Neural networks excel particularly in tasks like image and speech recognition, where the context and subtleties of data are paramount.

Deep learning is a subfield of machine learning that deals with neural networks having multiple layers, often referred to as deep neural networks. This is where the magic happens—helping in applications ranging from self-driving cars to advanced genomics. Deep learning requires massive amounts of data and computational power, which have become accessible with advancements in hardware and availability of large data sets. The way these models have developed to approach human-like reasoning is incredibly sophisticated, and each layer of the network hones in on specific detail, contributing towards building a result.

A fundamental concept in neural networks is backpropagation, a technique used for optimizing the weights of the neurons to minimize the error in predictions. What makes backpropagation significant is its iterative nature—adjusting weights based on the error calculated between the predicted output and actual output, thereby improving the model's accuracy over time. Every iteration is akin to learning from mistakes, not unlike how humans adjust their understanding based on past experiences.

The usefulness of machine learning and neural networks extends beyond typical commercial applications, diving deep into scientific research, healthcare, and even environmental sciences. In scientific research, machine learning processes have been applied for drug discovery and identifying new materials. In healthcare, it aids in diagnosing diseases from medical images with an accuracy that rivals human experts. Environmental sciences use these algorithms to build predictive models for climate change, species distribution, and resource utilization, allowing researchers to predict and plan for potential scenarios.

One of the notable challenges in machine learning and neural networks is understanding how decisions are made—often referred to as the "black-box problem." The complexity of these models makes it difficult to interpret why a model made a particular decision. This transparency issue is significant in critical areas where understanding the basis behind a decision is crucial, such as healthcare or autonomous driving. Researchers are actively working on explainable AI (XAI) to alleviate these concerns, aiming for models that both perform well and provide insight into their decision-making processes.

Another challenge is the ethics of data usage, which involves both privacy concerns and potential bias. Machine learning models are only as good as the data they are trained on. If the data contains biases, the models might learn and perpetuate those biases, which is particularly concerning in applications like facial recognition and hiring processes. Hence, there is a growing need for rigorous data preprocessing, inclusion of diverse datasets, and continuous monitoring to ensure fairness and inclusivity.

The scalability of these neural networks is astounding; when fed with parallel computing resources and optimized algorithms, they can handle immense datasets. This scalability has been pivotal because the more data a neural network is exposed to, the more nuanced its understanding becomes. However, as we reach for even more advanced applications, computational limits push the boundaries of what can be processed efficiently, highlighting the need for novel computational architectures and algorithms.

With these foundational building blocks, the interplay of machine learning and neural networks continues to inspire future innovations. As we delve deeper into the AI landscape, these technologies will not only drive the disruptive changes seen today but will also open new realms of possibility. They bridge the gap between insurmountable data and insightful knowledge, ensuring that the future of AI remains both promising and expansive.

As we journey forward in this unfolding chapter of technological evolution, the influence of machine learning and neural networks is akin to Newton's apple—offering a new view of the world and a toolset that extends our capabilities. Understanding the profound intricacies of these technologies unlocks a frontier for innovation and development that is limited only by our imagination.

AI Versus Human Intelligence

In the vast realm of technology, the comparison between artificial intelligence (AI) and human intelligence often sparks rich discussions, philosophical debates, and imaginative explorations. Our journey into the "Foundations of Artificial Intelligence" brings us face-to-face with one of the most fascinating aspects of AI—its comparison with the human mind. This intersection is where silicon meets synapse, and logic circuits confront intuition.

The idea that machines could exhibit intelligence was once purely science fiction. Yet, AI has evolved from theoretical algorithms into practical applications that affect every facet of life. It learns, adapts, and even reacts, but can it think as a human? This question sits at the heart of our inquiry. Although computers can process vast amounts of data at lightning speeds, their ability to 'understand' the nuances and context the way a human does is a different challenge altogether.

One significant aspect where AI often outshines human intelligence is in the consistency and speed of computation. Neural networks, for instance, can recognize patterns in data faster than any human ever could. Consider a medical AI system that scans thousands of radiological images to detect cancerous cells with greater accuracy than most human radiologists. Yet, despite its prowess, it operates within confines defined by its training data and lacks the emotional intelligence to truly understand the human experience of illness and vulnerability.

Human intelligence, on the other hand, thrives in adaptive reasoning and emotional depth. It excels in abstract thinking, creativity, and the ability to function with incomplete information. These nuanced capabilities are deeply rooted in our evolutionary journey. Humans have evolved to develop social paradigms and cultural understanding, which machines can only mimic but not inherently grasp. The way a human discerns emotion in a piece of art or interprets poetry involves layers of cognition and emotion that are still beyond AI's reach.

That said, AI continues to encroach domains once believed to be human-exclusive. Take creativity, for example. AI systems have composed symphonies, created visual art, and even written novels that resonate with people. The question arises: Can we equate these creations with those flowing from human minds? Does engaging randomly through algorithmic constraints equate to the curated spontaneous act of human creativity? Perhaps these questions lead us to redefine creativity altogether.

Moreover, AI and humans don't have to exist in opposition. Their combination can be complementary and symbiotic. AI can enhance human decision-making by augmenting our cognitive capabilities.

Consider AI-driven tools that help engineers design safer, smarter infrastructure by analyzing potential weaknesses that a human engineer might overlook. Similarly, in climate science, AI models predict weather patterns by processing climate data far exceeding human capacity. Together, they form a duo capable of greater accomplishments.

We must also be cautious in recognizing that AI, although powerful, can inherit biases from data it learns from. These biases can perpetuate discrimination and skew decision-making processes. Human oversight remains vital to guide AI's role positively in society. The human capacity for ethical reasoning and moral judgment is crucial for guiding AI development, ensuring these systems reflect and support societal values.

Besides, the intersection of AI and human intelligence invites questions about identity and the very essence of consciousness. Can a machine ever be self-aware or possess a consciousness resembling human experience? AI operates on logic and predefined algorithms, whereas human consciousness involves self-awareness, emotions, and subjective experiences. While AI can simulate decision-making and learning processes akin to human thought, a genuine consciousness remains in the realm of human experience.

As we explore the relationship between AI and human intelligence, it becomes evident that both have unique strengths and limitations. Where AI excels in speed, accuracy, and the ability to analyze vast datasets, human intelligence is unmatched in creativity, empathy, and ethical considerations. The future undoubtedly involves the seamless integration of both, where AI serves as a tool to augment and amplify human potential.

In the grand tapestry of technological evolution, AI stands as a powerful tool born of human ingenuity. As we move forward, the challenge is not about replacing human intelligence but enhancing it—

leveraging AI to push boundaries, solve complex problems, and create a harmonious coexistence. In doing so, we secure a future enriched by both human wisdom and the transformative potential of AI.

Chapter 3: Intersection of Quantum Computing and AI

The convergence of quantum computing and artificial intelligence represents a pivotal moment in technological evolution, promising to unlock a realm of possibilities previously thought unimaginable. By integrating the principles of quantum mechanics with AI capabilities, we're on the brink of transcending traditional computational limits and solving problems with unparalleled speed and efficiency. Quantum algorithms, with their potential to process vast datasets instantaneously, can revolutionize AI by expediting learning models and enhancing decision-making processes. This intersection doesn't just offer solutions but reframes challenges into opportunities, pushing the boundaries of what AI can achieve while inspiring a reimagined landscape of innovation. As we delve deeper into this synergy, we're poised to usher in an era where quantum-enhanced AI not only accelerates technological progress but also addresses some of the most pressing global challenges across sectors. This synthesis of technology invites us to envision a future where cognitive capabilities are not just expanded but fundamentally transformed, reshaping our understanding of intelligence itself.

Bridging Quantum Mechanics and AI

The interplay between quantum mechanics and artificial intelligence (AI) holds transformative potential in redefining computational capabilities. Traditionally, AI has harnessed the classical principles of computing, relying on bits as fundamental units of information. However, with the introduction of quantum mechanics into the computational domain, particularly through quantum computing, a paradigm shift is underway. This shift doesn't merely enhance existing processes but proposes fundamentally new approaches to problem-solving. Quantum mechanics, with its strange principles like superposition and entanglement, offers AI models unprecedented capabilities to tackle complex problems that were previously intractable under classical frameworks.

Quantum computing utilizes qubits, which, unlike classical bits, exist in multiple states simultaneously. A single qubit can be in a state representing both 0 and 1, thanks to the principle of superposition. This capability exponentially increases the computational power available to processes like neural networks and machine learning algorithms. When we combine this with AI's ability to adaptively learn from data and make predictions, new horizons open up. Tasks that require vast computational resources, such as large-scale optimization problems, can potentially be executed more efficiently using quantum-enhanced AI.

The concept of entanglement further enriches the toolkit available to AI researchers. Entangled qubits exhibit correlations that defy classical intuition, meaning the state of one qubit can instantly affect the state of another, regardless of distance. This characteristic could revolutionize AI, particularly in developing algorithms that require synchronized actions over distributed network nodes. The instantaneous connection facilitated by entanglement could lead to faster and more efficient training of neural networks, thereby advancing AI's learning capabilities.

While these theories are intellectually stimulating and full of promise, they are not without challenges. Implementing quantum mechanics into AI systems requires significant advancements in our understanding of both fields. Quantum algorithms must be conceived to utilize these unique properties efficiently. Here, the nascent field of quantum machine learning begins to take shape, where researchers endeavor to create algorithms that are inherently quantum, potentially unlocking efficiencies unreachable by classical algorithms.

The marriage of quantum computing and AI could lead to more nuanced decision-making processes, particularly in systems requiring high degrees of accuracy, such as autonomous vehicles or real-time data analysis in healthcare. The ability to process vast amounts of data quickly and discern patterns that would elude classical systems makes this integration highly desirable. Imagine a healthcare system where quantum-enhanced AI provides instantaneous and accurate diagnoses, even predicting potential health issues before they manifest visibly.

However, these possibilities walk hand in hand with intrinsic technical and philosophical challenges. Developing the infrastructure necessary to support quantum rights and computing poses complex questions. There's a need for reliable qubit systems, error correction mechanisms, and the development of user-friendly platforms that can effectively interface quantum systems with existing AI models. Ensuring these systems are both accessible and leverage the full potential of quantum mechanics is a significant hurdle.

Moreover, the ethical implications of integrating quantum mechanics with AI bear consideration. As these systems become more powerful and autonomous, defining the boundaries of AI's decision-making and ensuring fair and unbiased outcomes becomes crucial. The unpredictable nature of quantum processes introduces new dimensions to these debates, prompting discussions about privacy, control, and governance over quantum-enhanced AI decisions.

The exploration of quantum mechanics and AI also inspires questions about our cognitive limitations and the nature of intelligence itself. Can quantum-enabled AI reach an understanding of the universe that surpasses human comprehension? These questions straddle the line between science and philosophy, challenging us to rethink the limits of technology and the essence of human intellect.

In conclusion, the intersection of quantum mechanics and AI heralds a new era of technology that could reshape industries and alter the fabric of society. As we continue to bridge these two fields, the possibilities seem only limited by our imagination and the pace of scientific advancement. This junction could very well depict the dawn of a new understanding of intelligence, driven by quantum insights and AI ingenuity. In our quest to harness this intersection, we are writing the future of computational power, innovation, and intelligence.

Quantum Algorithms for AI

The intersection of quantum computing and artificial intelligence (AI) is a frontier full of potential that could reshape the very foundations of intelligent systems. It is particularly within the realm of quantum algorithms that AI stands to experience significant advancements. As these two transformative technologies evolve alongside one another, researchers and technologists are beginning to understand how quantum algorithms can become a powerful toolset for AI applications. Through unique computational paradigms, quantum computing offers a promise of enhanced processing capabilities that are far beyond the reach of classical systems.

Quantum algorithms are tailored to leverage the principles of quantum mechanics to conduct computations in ways that classical algorithms cannot. At the heart of this is the ability to operate on quantum bits, or qubits, which can exist in multiple states simultaneously due to the phenomena of superposition. This means quantum algorithms can process a vast amount of potential solutions at once, a feat that exponentially accelerates certain computations compared to classical algorithms. Furthermore, qubits can be entangled, allowing for information to be distributed across vast distances with instant correlation, introducing novel pathways for AI development that were previously unimaginable.

One of the most notable examples of quantum algorithms is Shor's algorithm, which is known for factoring large integers much more efficiently than the best-known algorithms running on classical computers. While its implications in cryptography are profound, its principles can be adapted to complex AI tasks, such as optimization problems that are critical in training machine learning models. Quantum optimization algorithms can explore the solution space of a problem more thoroughly and potentially find better solutions in significantly reduced timeframes.

Another significant quantum algorithm is Grover's algorithm, which brings a quadratic speed-up to unstructured search problems. In AI, this can be particularly beneficial for improving search-based tasks, such as query optimization and pattern recognition. The capability of Grover's algorithm to amplify the speed at which searches can be conducted aligns perfectly with the data-intensive nature of AI, opening up horizons of efficiency that classical algorithms struggle to compete with.

Moreover, quantum algorithms are being developed to address tasks in machine learning (ML), enhancing neural network operations. Quantum computing's ability to handle vast amounts of data and perform complex calculations efficiently could revolutionize how we train and deploy AI models. With techniques such as quantum-enhanced reinforcement learning, AI models can be trained on quantum computers to make faster and more nuanced decisions.

In a landscape where data volumes and complexity are rapidly increasing, quantum algorithms could potentially transform many subfields of AI, such as natural language processing (NLP), image and speech recognition, and decision-making algorithms. Quantum speed-up in handling these operations could alleviate the bottlenecks present in classical computing, especially when dealing with massive datasets, high-dimensional spaces, and nonlinear relationships that define AI complexity.

The integration of quantum algorithms into AI systems also brings opportunities for innovations that extend beyond just speed and power. It fosters the creation of new types of neural networks—and potentially, hybrid classical-quantum models—that can outperform traditional AI approaches in specific tasks. The blend of quantum probabilistic techniques with deep learning could lead to the development of more intuitive and seemingly graceful AI systems capable of understanding and interacting with complex environments in ways that mimic human-like intelligence more closely.

However, implementing quantum algorithms for AI is not without its challenges. The current nascent stage of quantum hardware, characterized by fragile qubits and error-prone operations, makes the realization of practical quantum computers a work in progress. Despite these hurdles, substantial efforts are being invested in developing quantum error correction and co-processing strategies that can facilitate more robust and error-resistant quantum computations, which are pivotal for realizing AI applications.

The ongoing research efforts are inspiring hope and optimism among scientists and technologists. Visionaries foresee a future where quantum-enabled AI applications solve problems that today seem insurmountable—such as drug discovery, personalized medicine, weather prediction, complex system simulations, and beyond. As we inch closer to achieving quantum supremacy, where quantum computers perform tasks unattainable by classical ones, quantum algorithms' potential impact on AI could instigate a paradigm shift in technology and infrastructure across various fields.

In conclusion, quantum algorithms are not just tools for enhancing AI; they're harbingers of a new era in intelligent computing. By blending quantum theory with AI's logical frameworks, we might be on the brink of breakthroughs in processing, understanding, and harnessing information. As research continues to make strides in overcoming current limitations, the collaborative rise of quantum computing and artificial intelligence promises to play a defining role in the technological advancements of the 21st century and beyond. Embracing these quantum possibilities could lead us toward a future where AI systems are more efficient, capable, and intricately intertwined with human endeavors.

Challenges and Opportunities

The intersection of quantum computing and artificial intelligence (AI) presents a unique confluence of challenges and opportunities that both excites and daunts researchers, technologists, and academics alike. A new era is brewing; one where the principles of quantum mechanics could radically enhance the capabilities of AI systems. However, snapping together these two complex and often unforgiving fields isn't without its hurdles. Yet, in overcoming these obstacles, the horizon of opportunities may stretch beyond what classical paradigms have permitted. Let's delve into the intricacies woven into this intersection and better understand what lies ahead.

To kick things off, one of the most significant challenges is the foundational difference in how classical and quantum computing paradigms operate. In classical systems, computational logic is rooted firmly in deterministic states—think 0s and 1s. Quantum systems, however, dance in the fluidity of superposition and entanglement, where qubits can exist in multiple states at once. This fundamental contrast means that integrating quantum computing into AI systems requires rethinking algorithms and architectures from the ground up. It's akin to redefining the rules of a game while it's being played.

Equally pivotal is the quagmire of *quantum decoherence.* Maintaining the delicate quantum states that provide computational superiority over classical systems is technically challenging and resource-intensive. Even fleeting environmental disturbances can disrupt these states, rendering computations moot. It's a massive engineering challenge, necessitating advances in error correction and the development of robust quantum hardware. These factors not only slow integration but pull heavily on research budgets and timelines.

Yet, despite these daunting technical challenges, the potential rewards are almost overwhelmingly compelling. Quantum-enabled AI promises to unlock previously inaccessible facets of problem-solving. Consider optimization problems in logistics, materials science, or cryptography. Quantum algorithms could relentlessly tackle these issues much faster and more accurately than classical algorithms. The prospect isn't just about squeezing more efficiency out of existing tasks; it fundamentally changes the scope of what we consider solvable.

One enticing opportunity lies in the realm of machine learning. Quantum computing could dramatically enhance the power of neural networks by increasing their capacity to process information. This serves to accelerate training times and enhance the accuracy of AI systems. For instance, think about the analysis of massive datasets where traditional methods fail or are inefficient. Quantum algorithms have the potential to read and interpret complex patterns and correlations hidden within data, enabling breakthroughs in fields from genomics to financial modeling.

Moreover, the integration of quantum computing could lead to radical improvements in AI's ability to learn and adapt. Quantum machine learning offers an avenue for developing new models that can operate in high-dimensional data spaces, potentially leading to insights that aren't just more detailed but more accurate. Conceiving an AI that autonomously improves upon itself daily isn't just speculation—in the quantum realm, it's a looming possibility.

However, where there are technological transformations, ethical considerations can't be left behind. As quantum AI stands on the cusp of revolutionizing industries, it raises pressing questions around accessibility, equity, and the potential for misuse. The democratization of this technology is critical to ensure it benefits a broad spectrum of society and not just a privileged minority with vast resources. Developing policies, frameworks, and checks that can evolve as fast as the technologies themselves is no small feat but an essential endeavor.

Securing global cooperation in setting standards and regulations also represents both a challenge and an opportunity. In an interconnected world, quantum AI can't function in isolation; it requires international collaboration to establish ethical guidelines and protect against inherent risks. Countries at different levels of technological advancement and ethical maturity need to work together, ensuring a balanced global perspective and playing field.

And then there's the opportunity for job creation and economic growth that's simply too significant to ignore. With new technologies come new industries and roles. As we delve deeper into mastering quantum AI, the demand for skilled professionals in quantum computing and AI will soar. This evolution promises not just economic dividends, but the chance to create thriving ecosystems of innovation and discovery.

In conclusion, the intersection of quantum computing and AI is a space ripe with both challenges and opportunities. While the journey is fraught with technical, ethical, and logistical hurdles, the potential payoffs make it a voyage well worth taking. The advancements made could redefine the boundaries of what's possible, crafting a future where not just industries change, but the very fabric of society itself is transformed. Embracing this intersection is not just about dreaming big—it's about preparing wisely and innovating relentlessly. The road may be winding and complex, but the destination promises a revolutionary leap forward in human capability and knowledge.

Chapter 4: Quantum Machine Learning

As we delve into the realm of quantum machine learning (QML), it's like standing at the junction where two titans of technology converge: quantum computing and artificial intelligence. At its heart, QML leverages the strange and powerful principles of quantum mechanics to enhance machine learning algorithms, ushering in a new era of computational possibilities. These quantum-enhanced methods promise to tackle complex problems and datasets that classical techniques struggle with, potentially solving tasks with exponential speed-ups. Imagine machine learning models that not only process patterns at unprecedented speeds but also draw insights from entangled states and superpositions. Though still in its nascent stages, the real-world applications—spanning finance, healthcare, and beyond—hint at a future where quantum speed and AI intellect meld seamlessly, driving transformative advancements across industries. The journey ahead is one of discovery, innovation, and, most excitingly, seeing how QML can reshape our understanding of intelligence and computational power.

Introduction to Quantum Machine Learning

As we venture into the realm of Quantum Machine Learning (QML), it's vital to appreciate the convergence of distinct intellectual domains that make this fusion possible. Quantum computing, with its roots in the peculiarities and profundities of quantum mechanics, and artificial intelligence, particularly its subfield of machine learning, come together to create a novel landscape poised to revolutionize computation. But what exactly is Quantum Machine Learning, and why does it represent such a pivotal advancement in the fields of technology and intelligence?

To start with, we need to understand the transformative power that quantum mechanics brings to computation. Traditional computers operate using bits as the fundamental unit of data—each bit representing zero or one. Quantum computers, however, utilize qubits, which harness the principles of superposition and entanglement. These concepts enable a qubit to exist in multiple states simultaneously, yielding exponential power in processing capabilities. This feature is crucial for addressing complex problems that classical computing struggles with, such as those involving vast datasets or intricate calculations requiring numerous variables.

Machine learning, a subset of artificial intelligence, drives many modern technologies, from recommendation algorithms to voice recognition systems. It allows machines to learn from and make predictions based on data, continually improving performance over time. Classical machine learning algorithms, however, face limitations, especially when dealing with highly complex or high-dimensional data structures. Enter quantum computing, offering a potential solution to these limitations by accelerating data processing and enhancing algorithmic efficiency.

Quantum Machine Learning stands at the crossroads of these advancements, opening new avenues for a slew of possibilities. By integrating quantum algorithms with machine learning models, we can envisage tackling challenges that previously seemed insurmountable, from optimizing large systems to simulating complex chemical processes. The marriage of these two fields is not merely an incremental step but represents a tenfold leap in capabilities, akin to the impact of transitioning from horse-drawn carriages to motor vehicles.

One immediate question that arises is: Why now? Why is QML gaining traction at this particular moment in time? The answer lies in the rapid advancements in both quantum technologies and AI methodologies. The last decade has witnessed significant breakthroughs in building quantum processors with more stable and reliable qubits. Concurrently, AI has seen tremendous growth, with machine learning models becoming ever more sophisticated, primarily due to advances in computing power and algorithmic design.

In this context, Quantum Machine Learning is more than just a novel concept. It's a vital tool empowering researchers to explore topics like quantum-enhanced neural networks, which you will delve into later in this chapter. These types of networks aim to capitalize on quantum computational advantages to create machine learning models that are not only faster but potentially more insightful.

However, it's not all rosy. The path toward fully realizing the potential of QML is fraught with challenges, both theoretical and practical. Quantum algorithms need careful design and host a distinct set of complications that classical algorithms do not encounter. The hardware itself is in an embryonic stage, with issues related to qubit coherence times and error rates yet to be fully resolved. Despite these hurdles, the academic and tech communities are relentlessly pushing forward, driven by the promise that QML holds.

Reflecting on the potential applications of Quantum Machine Learning, it's essential to highlight its broad implications across various domains. From optimizing supply chains for global enterprises to improving personalized medicine, QML's transformative potential cannot be overstated. It presents a unique opportunity to reimagine and reshape how we process information, unlocking new understandings of the natural world and complex systems. Quantum

Machine Learning is not just about speed; it's about creating a more connected, insightful, and smarter approach to problem-solving.

It is the dawn of a new era where the abstract lines between computing paradigms blur, and interdisciplinary collaboration becomes the cornerstone of progress. Researchers, technologists, and innovators from diverse fields—quantum physics, computer science, engineering, and beyond—are converging to tackle some of the most pressing and challenging problems of our time. The road ahead is rich with possibilities and rife with obstacles, making the journey exhilarating for scientists and engineers alike.

In sum, Quantum Machine Learning stands as a beacon of next-generation technology, embodying the potential to drive a paradigm shift in how we approach computation and learning. In subsequent chapters, we'll explore how these foundations built upon the theoretical underpinnings of quantum mechanics and machine learning come to life through practical implementations and case studies. The intersection of these technologies holds the promise of not just incremental advancements, but revolutionary transformations that will echo across industries and sciences, altering the landscape of what we previously deemed possible.

Quantum-enhanced Neural Networks

In the labyrinth of emerging technological paradigms, quantum-enhanced neural networks represent a fascinating intersection of quantum computing and artificial intelligence. As researchers strive to push the boundaries of what machines can achieve, the potential integration of quantum properties into neural networks hints at a seismic shift in computational capabilities. Understanding these hybrid systems requires not just a grasp of quantum mechanics, but an insight into how these principles can transform traditional AI architectures.

Neural networks, the backbone of today's most advanced AI systems, are modeled loosely on the human brain. They consist of layers of interconnected nodes or "neurons," which process data and enable machines to learn patterns and make decisions. The true power of neural networks lies in their ability to adaptively improve performance through training. However, the scalability and efficiency of these systems remain bottlenecked by the limitations inherent in classical computational architectures.

Enter quantum computing, which leverages principles such as superposition and entanglement to perform calculations that would be impossible, or at least infeasible, on classical computers. Quantum-enhanced neural networks aspire to integrate these quantum properties, potentially allowing for unprecedented efficiencies in processing power and the resolution of computational complexity. This hybrid approach could enable AI systems to solve problems with many more variables and configurations than classical systems currently manage.

The notion of quantum-enhanced neural networks begins with leveraging the concept of superposition. In classical computing, bits exist as either zero or one. In contrast, *qubits*, the building blocks of quantum computing, can exist in multiple states simultaneously. This allows quantum neural networks to process a vast array of inputs at once, refining and evaluating more data in parallel than would be possible classically. Such enhancements could significantly expedite processes like image recognition, natural language processing, or financial forecasting.

The element of entanglement, another hallmark of quantum mechanics, introduces unique pathways for quantum-enhanced neural architectures. Through entanglement, qubits can be interconnected in such a way that the state of one instantly influences the state of another, despite any physical distance between them. This feature may contribute to more robust model training and feature representation, potentially reducing the risk of overfitting—a common pitfall in machine learning where models become too tailored to specific datasets.

But beyond the theoretical, some significant practical challenges accompany the implementation of these cutting-edge networks. The environmental fragile states of qubits, error correction, and decoherence issues present persistent hurdles. Quantum-enhanced systems necessitate environments that support the delicate quantum states essential for their operation—something that is not trivial to achieve. Still, the race to develop quantum error correction techniques is progressing, as scientists and engineers explore highly promising paths to mitigate the challenges posed by quantum decoherence.

The marriage of neural networks with quantum technologies also involves a reimagining of algorithmic design. Unlike classical systems, where linearity often governs operations, quantum algorithms enable a shift out of traditional paradigms. Researchers are concentrating efforts on developing new kinds of algorithms that can exploit quantum speed-ups while integrating seamlessly with neural network structures. This could lead to the design of more generalized AI systems that can operate dynamically across a multitude of applications and environments.

While the full potential of quantum-enhanced neural networks remains an open field of exploration, the journey is already charting new territories in AI research. Initiatives by leading tech companies and academic institutions underscore the momentum and importance placed on this domain. IBM, Google, and startups specializing in quantum technologies are investing heavily in both theoretical research and practical applications, hoping to translate these advancements into real-world solutions.

As we forge ahead, ethical considerations also find a place in the discourse surrounding quantum-enhanced neural networks. The considerable power and adaptability of such systems demand a measured approach to their development and deployment. Issues of privacy, control over AI decision-making, and the potential for biases embedded within quantum-enhanced models must be critically addressed to ensure these technologies are aligned with foundational societal values.

The potential applications for quantum-enhanced neural networks span across nearly all sectors. In medicine, they could revolutionize drug discovery processes, potentially leading to expedited identification of novel treatment pathways. In finance, they might unlock new methods for risk assessment and fraud detection, offering unprecedented precision. In climate science, leveraging the power of quantum-enhanced models might allow us to better predict complex climate patterns, aiding in sustainable global planning.

While we are still in the early days, the inevitable fusion between neuroscience-inspired neural architectures and quantum computing's quantum leap in capabilities inspires both excitement and caution. Pioneers in this field are not only working to harness the raw computational power of quantum systems but are also steering their potential towards making these new neural designs practically applicable. How we navigate this landscape will ultimately define the impact quantum-enhanced neural networks have on the technology of tomorrow.

The journey towards fully-functioning quantum-enhanced neural networks echoes with both promise and complexity. As they become an integral part of the AI ecosystems, they may redefine expectations and possibilities, ultimately expanding humanity's ability to tackle large-scale challenges. Echoing the bold spirit of exploration that characterizes our technological age, these networks could unlock extraordinary frontiers and dare us to dream even bigger.

Real-world Applications and Case Studies

Quantum machine learning (QML) is on the brink of transforming how industries operate, providing revolutionary approaches to problem-solving across a diverse range of sectors. As the synergy between quantum computing and machine learning crystallizes, it's paving the way for innovative applications that promise not only efficiency but also previously unattainable insights.

Consider the pharmaceutical industry, where drug discovery is a costly and time-consuming process. Traditional methods often involve sifting through millions of compounds, requiring computational power that classical computers struggle to deliver efficiently. Quantum algorithms, however, can handle and interpret complex molecular and atomic interactions more precisely. Case in point: companies like IBM and Google are exploring quantum computing to simulate molecular structures, potentially reducing the time and cost involved in bringing new drugs to market.

Similarly, in the realm of material science, quantum machine learning holds the promise of identifying and developing new materials with unprecedented properties. Industries such as energy, telecommunications, and electronics stand to benefit from the development of superconductors or materials with higher efficiency rates discovered through quantum-enhanced algorithms. Armed with the ability to process combinations and their interactions at subatomic levels, materials predicted to be both stronger and lighter could transform infrastructure and technology.

Financial services, ever at the forefront of technological adoption, are also diving into QML. Quantum algorithms for portfolio optimization allow financial institutions to process data at unprecedented speeds, yielding more accurate predictions and minimizing risk. Take JPMorgan Chase, which has been active in applying quantum computing to assess risks and manage assets, aiming to offer quicker and more insightful analyses to their clients.

Another critical area where QML is making waves is logistics and transportation. Companies like Volkswagen are using quantum machine learning to optimize traffic flow, potentially reducing congestion and emissions. QML algorithms can process vast amounts of data, providing real-time insights and forecasts that are crucial for efficient network management. Imagine a future where traffic jams are a relic of the past, thanks to timely interventions powered by quantum insights.

The promise of quantum machine learning extends into the realm of cybersecurity too. As global threats become more sophisticated, the need for advanced defensive measures grows. Traditional encryption methods are increasingly vulnerable, prompting a shift towards quantum cryptography, which provides theoretically unbreakable encryption. Integrating QML into cybersecurity strategies can lead to rapid identification of threats and adaptive measures, safeguarding sensitive information like never before.

Moving into healthcare, QML's potential for patient diagnosis and treatment personalization cannot be overstated. AI-driven technologies, enhanced by quantum computing, can deliver rapid data analysis and pattern recognition, offering medical professionals the ability to personalize treatments based on genetic data. This transformational approach not only promises improved health outcomes but also a more efficient healthcare system, opening doors to treatments tailored to individual genetic profiles.

Precision agriculture is another promising domain. With QML, farmers can predict weather patterns, optimize crop yields, and manage resources efficiently, all by analyzing complex environmental data.

Quantum-enhanced algorithms can process high volumes of data from soil sensors, satellites, and climate models, offering precise recommendations for sustainable farming practices.

In terms of AI itself, QML represents a leap forward in the development of more powerful neural networks and learning models. By overcoming the limitations of classical machine learning approaches, researchers can develop systems capable of solving significantly more complex problems. For example, Google's efforts with their Quantum AI team illustrate how combining quantum computing with deep learning algorithms can unlock new opportunities, from improving natural language processing to creating more sophisticated AI systems capable of more complex tasks.

Yet, it's not just the commercial and practical applications where QML shines. Cultural sectors like art and music are beginning to tap into the potential of QML to create new forms of digital art and compositions. By harnessing the randomness and probability inherent in quantum systems, artists can explore novel pathways unimagined in the classical world, producing pieces that challenge traditional boundaries and captivate audiences with innovation.

Despite the existing promise, however, real-world deployment of QML is not without its challenges. Scalability, error rates, and the current limited number of qubits are significant hurdles in its widespread adoption. Yet, with continuous advancements and research, these obstacles are slowly being surmounted, promising a future where QML's impact becomes ubiquitous.

In essence, the real-world applications and case studies of quantum machine learning highlight an impending technological leap. As industries continue to integrate these advancements, the potential to reshape our world positively through more intelligent, more efficient, and highly tailored solutions becomes not just a possibility but a progressively imminent reality. Every leap in quantum machine learning suggests a cascading impact across various sectors, hinting at an era where technology and human ingenuity harmoniously converge for remarkable achievements.

Chapter 5:
Quantum AI for Data Analysis

In the rapidly evolving landscape of technology, Quantum AI is emerging as a formidable ally in revolutionizing data analysis, transforming how we tackle complex datasets. At its core, Quantum AI melds quantum computing's unparalleled processing prowess with advanced artificial intelligence algorithms to sift through vast oceans of information, offering insights previously out of reach with classical methods. This dynamic synergy enables the handling of big data with unprecedented efficiency, particularly in analyzing intricate systems that demand nuanced solutions. By uncovering patterns in labyrinthine datasets, Quantum AI not only accelerates decision-making processes but also paves the way for innovations across various sectors, from finance to healthcare. Its potential to redefine data analytics heralds a new era where precision and speed are no longer mutually exclusive, ultimately equipping us with tools to address the world's most pressing challenges with agility and foresight.

Quantum Tools for Big Data

As we navigate the throes of the digital age, the surge of big data creates both formidable challenges and unprecedented opportunities. Here, quantum computing emerges as a transformative force, redefining how we approach massive datasets. With its unparalleled capabilities, quantum AI brings forth tools that shatter the constraints of classical systems. By leveraging the unique principles of superposition and entanglement, quantum data analysis can process a wealth of information with incredible efficiency, offering insights that were once out of reach. These tools enable us to decipher complex patterns and correlations within large-scale data, propelling industries toward solutions that enhance innovation and efficacy. As we integrate these quantum tools, we're not just inching closer to solving big data dilemmas; we're laying the groundwork for a future where data-driven decisions pave the way for transformative advancements across multiple sectors.

Handling Large-scale Data with QML Amid the digital revolution, the proliferation of data presents both a challenge and an opportunity. As datasets grow in size and complexity, traditional data analysis tools often struggle to keep up. This is where quantum machine learning (QML) steps in, offering potent solutions for handling large-scale data. Harnessing the unique properties of quantum computing, QML has the potential to transform the landscape of data analysis, allowing for unprecedented efficiency and insight.

QML stands at the intersection of quantum computing and machine learning, combining the strengths of both fields. Quantum computers leverage phenomena like superposition and entanglement, enabling them to process information in ways classical computers can't. For large-scale data, this means QML can explore intricate datasets more swiftly and accurately than their classical counterparts. By applying quantum algorithms to machine learning models, QML can potentially unlock new patterns and relationships obscured by sheer data volume.

At the heart of QML's capability to manage large datasets is its innate parallelism. Quantum computers don't operate in the binary constraints of classical systems. Instead, they utilize qubits, which exist in multiple states simultaneously. This multi-state capacity facilitates the simultaneous analysis of vast data points, accelerating computational processes that would take classical systems significantly longer to complete. Such speed and efficiency are critical when tackling big data challenges across industries.

Consider, for instance, the unprecedented growth of data in fields like genomics and climate science. Here, every second counts, and the need to sift through terabytes of data quickly and efficiently can mean the difference between groundbreaking discovery and missed opportunity. QML can scale to these challenges, providing scientists and researchers with tools to make more informed decisions faster than ever before.

But handling large-scale data isn't just about speed. It's about the precision of insights. QML not only processes data rapidly but also enhances the accuracy of the models built upon this data. Quantum algorithms can improve machine learning tasks such as classification, clustering, and regression, often exceeding the accuracy attainable through classical means. This is particularly beneficial when translating data patterns into actionable intelligence where precision is paramount.

Moreover, the scalability of QML promises significant breakthroughs in fields that require real-time data processing and decision-making. Industries like finance and healthcare, where speed and accuracy are critical, can benefit tremendously. Quantum systems can potentially provide real-time analytics for stock markets, fraud detection, or patient diagnosis, all of which hinge on the rapid, reliable processing of large datasets.

The integration of QML in handling massive data doesn't come without challenges. Current quantum systems face issues related to noise and error rates, which can affect accuracy. Additionally, the field is still in its infancy, with continuous research required to refine quantum algorithms and hardware to realize their full potential. As these technological hurdles are addressed, the promise of QML in revolutionizing data analysis becomes increasingly tangible.

Collaboration will be crucial to overcoming these challenges. Partnerships between academia, industry, and government can drive the innovation needed to advance QML technologies. As stakeholders pool resources and expertise, they can accelerate the development of more robust quantum systems and algorithms tailored to manage large-scale data effectively.

As we look to the future, the implications of integrating QML for large-scale data handling are profound. Not only does it promise dramatic improvements in how we analyze data, but it also paves the way for novel applications in AI and beyond. By unlocking the potential of quantum-enhanced data analysis, we stand on the precipice of a new era in technology where data-driven insights shape our understanding of the world around us more deeply and effectively than ever before.

Analyzing Complex Systems

In the ever-evolving landscape of technology, understanding and analyzing complex systems represent a critical undertaking. Quantum AI, heralded as a groundbreaking synthesis between quantum computing and artificial intelligence, is uniquely positioned to address this challenge. Complex systems, by their nature, exhibit a level of unpredictability and interconnectivity that classical computational methods struggle to decode. Quantum AI introduces a paradigm that not only embraces complexity but thrives on its intricacies, making it an ideal tool for a broad spectrum of applications.

At the heart of analyzing complex systems is the concept of interconnectedness. Traditional models often reduce systems into isolated components in an attempt to simplify analysis. However, this reductionist approach can miss the subtle nuances that are crucial for a true understanding of the system's behavior. Quantum AI, with its roots deep in the principles of quantum entanglement and superposition, offers a revolutionary way to process interdependent data points simultaneously. This ability allows for a holistic view that is essential when dealing with systems as vast and varied as ecosystems, urban infrastructure, or even the human brain.

The potential for Quantum AI to transcend classical computational limitations is profound. Consider the task of modeling climate patterns—a quintessential complex system continually influenced by innumerable variables. Quantum algorithms can process multiple environmental variables at once, providing insights with unprecedented fidelity and speed. This isn't just an incremental improvement; it's a transformative leap that can lead to more accurate predictions, enabling societies to better prepare for and mitigate climate-related disasters.

Another domain ripe for Quantum AI analysis is social systems. In these systems, human behavior interacts in intricate and often unpredictable ways. Understanding these dynamics is vital for improving urban planning, public policy, and even economic forecasting. Quantum AI's ability to process and learn from vast quantities of data makes it an extraordinary tool for uncovering patterns and correlations that might otherwise remain hidden. Such insights could support the development of more effective strategies to address social issues—from healthcare distribution to crime prevention—by better understanding the root causes and potential interventions.

In finance, the complexity of market ecosystems is magnified by their global scale and interconnectedness. Traders and analysts have long sought methods to decode market trends, mitigate risks, and make informed decisions. Quantum AI offers a sophisticated approach to modeling financial markets by considering the myriad interactions between economic indicators, company performance, and geopolitical events. With its enhanced computational power, Quantum AI can provide a clearer picture of financial landscapes, leading to better risk assessments and possibly unveiling new investment opportunities.

Healthcare systems, with all their interconnected components ranging from patient records to genetic information, also stand to benefit from Quantum AI. Analyzing patient data with a quantum-enhanced lens could identify new correlations between symptoms and diseases, tailor treatments to individual genetic profiles, and streamline operations management. Quantum AI's power to manage large datasets efficiently and accurately may well lead to more personalized and effective healthcare solutions.

Even in creative industries, analyzing complex systems plays a crucial role. Whether examining the intricate network of influences shaping contemporary music or understanding the multifaceted interplay of elements in digital art, Quantum AI paves the way for creative analysis and innovation. By offering new insights into creative processes and outputs, Quantum AI may one day help artists uncover novel forms of expression or anticipate emerging trends.

The road to fully integrating Quantum AI in analyzing complex systems isn't without its challenges. Quantum systems, while powerful, are still maturing. Robust error correction and the scalability of quantum hardware are just a few hurdles needing refinement before these systems can reach their full potential. However, as breakthroughs continue and technology progresses, the marriage between quantum computing and AI promises to make analyzing complex systems not just feasible, but remarkably efficient.

In conclusion, the fusion of quantum mechanics with artificial intelligence heralds a new era for analyzing complex systems. The potential applications are as diverse as the fields they touch—ranging from environmental science to financial modeling, social systems to healthcare, and beyond. The methodical exploration of Quantum AI for complex systems analysis could not only illuminate hidden dynamics within such systems but also offer solutions to some of the most critical challenges of our time. Embracing this technology with curiosity and a spirit of innovation will undoubtedly unlock possibilities that we are just beginning to imagine.

Chapter 6: Cybersecurity in the Quantum AI Era

As we navigate the complex landscape of the Quantum AI Era, cybersecurity evolves into a pivotal arena, where traditional defenses face unprecedented challenges and electrifying possibilities. The tremendous capabilities of quantum computing threaten to render current encryption methods obsolete, ushering in the need for quantum-powered cryptography that can withstand such formidable computational power. Meanwhile, AI's role in cyber defense transforms threat detection and response strategies, offering adaptive, intelligent systems capable of analyzing vast landscapes of digital data in real-time to combat emerging cyber threats. The synergy between quantum technologies and AI doesn't just amplify the potential for cyber resilience—it reshapes our entire strategic approach to safeguarding data in an increasingly interconnected world. In this dynamic interplay between offense and defense, we find inspiration in the opportunity to fortify our digital realms while navigating the challenges of an ever-evolving technological frontier.

Quantum-powered Cryptography

As quantum computing continues to redefine the contours of technology, one of its most impactful domains is cryptography. Cryptography, the ancient art of secure communication in the presence of adversaries, stands on the cusp of a revolution. The emergence of quantum-powered cryptography isn't just a gradual evolution—it's a monumental leap that could alter the very foundations of digital security. Quantum cryptography promises unparalleled security capabilities, rooted in the principles of quantum mechanics, that classical methods simply can't offer. But what makes this quantum-driven approach to cryptography so revolutionary? To answer this, we must first dive into the unique attributes of quantum mechanics that redefine our traditional cryptographic paradigms.

At the heart of quantum mechanics lies the principle of superposition, where particles like electrons exist in multiple states at once until measured. This phenomenon is deftly harnessed by quantum cryptography through its encryption keys. Unlike classical keys, which are sequences of bits that attackers can feasibly break with enough computational power, quantum keys have the potential to remain perpetually secure. The use of quantum bits, or qubits, allows for the creation of Quantum Key Distribution (QKD) protocols that are theoretically impervious to eavesdropping. Because any attempt to observe quantum states alters them, eavesdroppers can be detected with certainty, thus rendering the communication immediately suspect.

An embodiment of QKD is the famous BB84 protocol, named after its inventors Charles Bennett and Gilles Brassard. The BB84 protocol uses polarization states of photons to encode bits, ensuring that any interception results in detectable disruptions. In practical terms, this ensures that communications are as secure as nature permits. Techniques like BB84 represent a shift from relying on computational difficulty as a basis for security to leveraging the fundamental, immutable characteristics of physics. This shift highlights the transformative potential of quantum cryptography: instead of trying to make decryption harder for an attacker, quantum cryptography makes it evident when a system is under attack.

The implications of quantum-powered cryptography stretch beyond just economic sectors or governments. Every digital transaction, from personal emails to billion-dollar trades, could one day rely on quantum-secured channels. This newfound security potential, while exhilarating, also poses substantial challenges. Current computational practices, deeply entrenched and widespread, will require significant overhauls. Transitioning to quantum-secure methods demands not only technological upgrades but also a paradigmatic shift in how organizations perceive and implement their security strategies.

However, lest we forget, the flip side of this quantum advancement lies in the potential threat to classical cryptographic algorithms. Quantum computers, equipped with algorithms like Shor's, have the prowess to decipher widely-used encryption schemes like RSA exponentially faster than classical systems. As per RSA's traditional framework, these encryption schemes depend heavily on the difficulty of factoring large numbers. Shor's quantum algorithm, with its capacity to solve such problems efficiently, poses an existential threat to RSA and alike.

This dual role of quantum technology—both as a guardian through quantum cryptography and a possible rival through quantum decryption abilities—crafts a compelling narrative for the future of cybersecurity. Cryptographers and security experts must prioritize the development of post-quantum cryptographic solutions, anticipating a future where quantum computers are ubiquitous. Even now, researchers are actively working on algorithms that can withstand quantum attacks, seeking to integrate post-quantum cryptographic standards into global security architectures.

Moreover, the implementation of quantum-powered cryptography is more than just securing private transactions. It extends into maintaining the integrity of diplomatic communications and national securities. Governments worldwide view the mastery of quantum cryptography as a strategic priority, understanding its potential to alter geopolitical balances. As nations race toward quantum supremacy, those who effectively integrate quantum cryptography could set new standards for international security protocols.

Nevertheless, deploying quantum cryptography at scale is fraught with challenges. While theoretical models like QKD promise high levels of security, practical implementations still confront hurdles, among them are reliable and efficient qubit management and error correction. Existing infrastructure must adapt, often requiring quantum networks that can ensure fault tolerance and interoperability across vast distances. Striking a balance between innovation and existing infrastructure is crucial for harnessing quantum cryptography's potential without succumbing to inefficiencies or vulnerabilities during the transition phase.

The allure of quantum-powered cryptography remains its promise of an almost unassailable layer of security in an increasingly interconnected world. Whether it's through the intricate dance of qubits defying conventional logic or the profound implications for global security, what lies ahead is a horizon filled with both promise and responsibility. As we stand at the threshold of this quantum era, the role of quantum cryptography is not merely to secure data, but to usher in a new era of trust and transparency across the digital world.

To conclude, amidst the complexities that come with quantum advancements, the potential for revolutions in cryptographic techniques is immense. While the journey towards a quantum-secure world is fraught with challenges, it's also brimming with opportunities for innovation. With careful, strategic advancements, quantum-powered cryptography is set to become the linchpin of modern cybersecurity, transforming how we safeguard our digital lives and, in turn, shaping the very fabric of the global information landscape. As industries, policymakers, and technologists work in concert, the true promise of quantum cryptography can be realized, setting a new gold standard for security in the digital age.

AI in Cyber Defense

In the ever-evolving landscape of cybersecurity, keeping pace with the rapid advancements in technology is both a challenge and necessity. We're entering an era where AI-driven cyber defenses are poised to transform how we protect digital infrastructure. As quantum computing continues to advance, the potential for artificial intelligence to revolutionize cybersecurity becomes more pronounced. Traditional methods of defense are no longer sufficient against increasingly sophisticated threats. AI in cyber defense is not merely a luxury but a pivotal shift that stands at the forefront of modern cybersecurity strategies.

AI's role in cyber defense lies majorly in its ability to process vast amounts of data. It can detect patterns and anomalies far beyond human capabilities. Cyber threats often come with subtle changes, making them hard to identify using conventional algorithms. But with AI, we can create dynamic models that continuously learn and evolve, identifying risks before they materialize into cyber attacks. This proactive approach significantly increases the odds of preempting breaches and thwarting attacks.

The brainpower of AI in cybersecurity resides in machine learning, which enables systems to learn from historical data. AI systems can differentiate between normal and malicious activities, creating a baseline of 'normal' behavior for networks. When AI systems spot deviations from this baseline, they can flag potential threats. Unlike traditional systems that rely on predefined rules, AI approaches blend flexibility and adaptability. They're not only able to recognize known threats but also detect what has yet to be identified, such as zero-day vulnerabilities.

Moreover, AI has proven to be a formidable ally in the realm of real-time threat analysis. Traditionally, human cybersecurity experts faced challenges in processing and analyzing the growing number of alerts. The sheer volume of data necessitated automated solutions. By implementing AI, security teams can automate the analysis and prioritization of threats. AI algorithms filter through noise and present actionable insights, allowing professionals to focus on responding to critical threats rather than sifting through potentially benign data.

An exciting development in AI-driven cyber defense is the emergence of AI-enhanced threat intelligence platforms. These platforms analyze threat data across entire industries, compiling vast libraries of threat activities and malicious behavior patterns. AI can test these scenarios against company networks, simulating attack vectors and defenses in a controlled environment. The feedback from such exercises fortifies defenses in real-time, enabling organizations to handle real-world cyber threats with increased confidence.

Incorporating AI into cybersecurity strategies, however, requires cautious implementation. One significant challenge is avoiding over-reliance on AI systems. These advanced systems need human oversight to ensure accuracy and accountability. AI can augment human capabilities but doesn't replace the intuitive judgment and creativity of skilled cybersecurity professionals. Therefore, a hybrid approach combining AI's capabilities with human insight remains crucial.

Furthermore, as beneficial as AI is for defense, it can also empower attackers. Malicious actors are already developing AI systems to design smarter malware and conduct more effective phishing attacks. The very technology that promises to secure us can be weaponized. This dual-edged nature of AI makes it imperative for security experts to stay ahead of potential threats, constantly learning and adapting AI models to anticipate and mitigate future challenges.

Another crucial component of AI in cyber defense is the development of AI-oriented cybersecurity talent. Cultivating expertise in both AI and cybersecurity will be vital as we continue to integrate these systems. Educational institutions and organizations must focus on interdisciplinary training that combines AI technology with cybersecurity knowledge. This specialized workforce will be capable of developing and maintaining robust AI-driven defenses, ensuring that we remain one step ahead of adversaries.

Moreover, partnerships across industries and nations are essential to address the global AI-cybersecurity landscape. Cyber threats do not recognize borders, and so collaboration becomes a necessity. Sharing data and intelligence across industries and governments can develop more comprehensive AI-driven defense strategies. Such cooperation brings diverse perspectives to the table, enhancing the overall defense mechanism and enabling faster, coordinated responses to cyber threats.

Looking ahead, the integration of quantum computing with AI may unlock unprecedented cyber defense capabilities. Quantum AI could potentially conduct large-scale data analysis at previously unimaginable speeds, further advancing both detection and response mechanisms. As quantum computing becomes more prevalent, the synergy between these two fields will redefine cybersecurity paradigms. Quantum AI could evolve the very concept of defense, incorporating innovative quantum algorithms designed for cybersecurity applications.

The path forward is filled with challenges, but the prospects are equally promising. As AI continues to evolve, so too will our capacity to secure digital infrastructures, safeguard sensitive information, and ensure the integrity of our interconnected world. By leveraging AI in cyber defense, we're not just reacting to cyber threats; we're stepping into a future where potential threats are addressed with foresight and precision, ensuring a safer digital ecosystem for all.

Combating Emerging Cyber Threats

In the dawn of the quantum AI era, cybersecurity is confronting an unprecedented pivot. The coupling of quantum computing and artificial intelligence presents both profound opportunities and stark challenges. While these technologies can fortify defenses, they also empower adversaries with sophisticated tools. Our digital world, once protected by complex mathematical puzzles, now faces the threat of these puzzles being unraveled at speeds inconceivable by classical computers.

The landscape of cyber threats is evolving rapidly, with quantum computing's imminent ability to crack widely-used encryption methods. Traditional encryption practices like RSA and ECC rely on the difficulty of factorizing large numbers, a task quantum computers can handle with ease through Shor's algorithm. As these quantum capabilities emerge, they endanger the very foundations of our current cybersecurity frameworks. Organizations and governments, aware of this impending threat, are racing to develop quantum-resistant algorithms that can withstand the computational prowess of future quantum machines.

Artificial intelligence, meanwhile, plays a dual role in this saga. On one hand, AI systems enhance security by predicting threats through pattern recognition, anomaly detection, and real-time data analysis. They can parse through vast quantities of data, spotting potential threats that human analysts might miss. But on the other hand, these same AI technologies can be exploited to devise sophisticated cyberattacks. AI's capacity for mimicry and adaptation allows malicious actors to create phishing attacks that are indistinguishable from genuine correspondence or to disrupt systems by learning defensive patterns and exploiting them.

As we strategize to combat these emerging threats, collaboration becomes pivotal. Multifaceted teams comprising cryptographers, computer scientists, and AI specialists are working together to innovate defensive mechanisms. Quantum cryptography, leveraging principles of quantum mechanics, such as quantum key distribution (QKD), offers new promise. By using the peculiarities of quantum states to transmit data, QKD provides a theoretically unbreakable method of securing information, as any attempt to intercept keys alters their state, signaling potential breaches.

The integration of AI with quantum cryptography introduces a powerful defensive shield. AI can monitor cryptographic systems, ensuring keys remain uncompromised and enhancing transaction security. These systems, with their ability to learn and adapt, can offer dynamic security solutions that evolve with the nature of threats. This adaptability is crucial in a world where cyberthreats morph rapidly, and static defenses quickly become obsolete.

Furthermore, pioneering efforts in privacy-preserving machine learning aim to safeguard sensitive data against quantum-enabled breaches. Techniques such as homomorphic encryption allow computations on encrypted data without needing to decrypt it first. These methods protect the confidentiality and integrity of data, reducing vulnerabilities that adversaries might exploit. By integrating these techniques into AI algorithms, we can bolster systems against emerging quantum threats.

Understanding the adversaries employing these technologies is equally crucial. Cybercriminals and hostile entities are quick to exploit cutting-edge developments, and the complex interplay between AI and quantum technologies offers them a fertile battleground. To preempt these threats, developing a robust cybersecurity culture that extends from leadership to the grassroots level is imperative. Companies and governments are prioritizing cybersecurity education, emphasizing the importance of protocols and the significance of timely incident reporting.

Looking beyond the technological and strategic realms, ethical considerations loom large over these advancements. The power granted by quantum AI can easily slide towards malevolent applications if not carefully regulated and monitored. A framework of policies and ethics in cybersecurity, ensuring that advancements serve the broader good, must underpin the technological race. This requires international cooperation, where nations share insights and policies, building a united front against global cyberthreats.

The transition to a quantum AI-secure future won't be seamless. It demands substantial investment in research and the development of new standards that integrate seamlessly into existing infrastructures. Organizations need to conduct regular assessments of their systems, akin to thorough medical checkups, ensuring resilience against both active and emerging vulnerabilities. They must also keep a vigilant eye on advancements from adversaries, swiftly adapting to new tactics as they arise.

Nonetheless, amid the challenges, this era grants us an opportunity to redefine cybersecurity. By adopting proactive cybersecurity postures and fostering innovation within ethical boundaries, we can pave the way for a digitally secure future. This involves not just viewing quantum AI as tools to combat threats, but as a new paradigm to understand and outpace them. By cultivating a future-oriented mindset, prioritizing agile methodologies, and fostering interdisciplinary collaboration, we pivot from being reactive to predictive, from being apprehensive to being empowered.

As we continue to unravel the complex tapestry of quantum AI and its implications for cybersecurity, the question isn't merely how to combat these emerging threats, but how to harness quantum AI's full potential to create a resilient digital ecosystem. Each breakthrough not only unveils new dimensions of potential peril but also provides us with innovative pathways to forge robust defenses. In overcoming these challenges, we don't just shield our digital assets; we transform the future of cyberspace into one marked by trust, security, and innumerable possibilities.

Chapter 7: Revolutionizing Healthcare with Quantum AI

The convergence of quantum computing and artificial intelligence is poised to disrupt the healthcare industry in unprecedented ways. As the intricacies of human biology unfold, quantum AI steps in, promising to unlock new dimensions in medical research and innovation. With its unparalleled computational power, quantum AI accelerates the discovery of complex patterns in genomic data, potentially unraveling mysteries of diseases at a molecular level, thus paving the way for breakthroughs in medical research. Moreover, the fusion of quantum computing and AI facilitates personalized medicine by tailoring treatment plans to the genetic makeup of individuals, enhancing the precision and effectiveness of medical interventions. The sphere of AI-driven diagnosis and treatment also witnesses transformation, as quantum algorithms analyze vast patient datasets with extraordinary speed and accuracy. This not only optimizes treatment protocols but also forecasts disease progression, providing a proactive rather than reactive approach to healthcare. As these technologies continue to evolve, they herald a new era of medical possibilities, fundamentally changing how diseases are understood, diagnosed, and treated, promising a healthier future for humanity.

Breakthroughs in Medical Research

In the ever-evolving landscape of healthcare, the marriage of quantum computing and artificial intelligence (AI) heralds a new era—a transformative revolution that promises unprecedented breakthroughs in medical research. At the heart of this transformation lies the potential of quantum AI to redefine diagnostics, therapeutics, and our understanding of complex biological systems.

Traditional computing methods have reached significant milestones in processing medical data, but they often hit a ceiling when it comes to decoding the intricacies of human biology. Enter quantum computing, with its ability to handle enormous datasets and perform complex computations that were once relegated to science fiction. By leveraging principles such as superposition and entanglement, quantum computers can process multiple possibilities simultaneously, thus accelerating research timelines.

One area where quantum AI is poised to make a distinct mark is in genomics. The human genome contains over three billion base pairs—an astronomical challenge for classical computation. Quantum computers, however, can potentially analyze genetic sequences in ways we couldn't have imagined before, leading to faster identification of genetic mutations that cause diseases. Imagine the possibilities: pinpointing rare genetic disorders within hours, sparking new lines of investigative research, and customizing treatment plans on a patient-by-patient basis. This isn't just revolutionizing genomics; it's rewriting the rules of personalized medicine.

Additionally, in fields such as drug discovery, quantum AI is an emerging powerhouse. The traditional drug development process is notoriously lengthy and expensive, often taking over a decade and billions of dollars to bring a single drug to market. Quantum computing can simulate molecular interactions at an unprecedented level of detail, rapidly narrowing down potential compounds that could yield effective treatments. By drastically reducing trial and error, quantum AI not only speeds up the development cycle but also increases the chances of finding groundbreaking cures for diseases previously considered untreatable.

Consider oncology, a field where precision is paramount and time is of the essence. Quantum AI could redefine cancer treatment by offering a deeper understanding of tumor biology, identifying novel therapeutic targets, and optimizing treatment protocols. By integrating AI-driven algorithms with quantum-enhanced data analysis, researchers can devise treatment plans that adapt in real-time to how a patient's cancer behaves, offering a truly personalized and dynamic approach.

Quantum AI's impact isn't limited to the cellular level. It extends to the macro realm of public health as well. Epidemics pose one of the most significant challenges of our age, where speed and accuracy in response can dictate life-or-death outcomes. Through enhanced predictive models, quantum AI could help in anticipating the spread of infectious diseases, evaluating intervention strategies, and optimizing resource allocation. The COVID-19 pandemic underscored the critical need for such advanced tools, showing that every moment saved can mean lives saved.

Another fascinating application lies in the realm of neurological research. The human brain is an incredibly complex organ, with billions of neurons firing synchronously—a system classical computers struggle to fully comprehend. Quantum AI could unlock new insights into neurological disorders like Alzheimer's or Parkinson's by simulating these neural networks at a quantum level. This could lead to technologies that not only predict the onset of such diseases but also offer avenues for early intervention.

Moreover, tackling the ethical challenges of medical research becomes intrinsically linked to quantum AI's development. Healthcare professionals and researchers are tasked with balancing rapid technological progress with the ethical imperative to do no harm. Quantum AI amplifies this balance by providing more accurate results, reducing the likelihood of systematic biases that can emerge from datasets, and offering solutions that accommodate diverse patient populations. This capacity to enhance equity in healthcare cannot be overstated.

The collaboration between healthcare professionals and tech innovators forms the backbone of these breakthroughs in medical research. The interdisciplinary synergy catalyzes growth and exploration in uncharted territories of medicine. Biomedical researchers now work hand-in-hand with quantum physicists and AI specialists to develop applications uniquely suited to the needs of the medical community.

The potential of quantum AI in public health policies and healthcare systems governance is equally transformative. By offering data-driven insights into healthcare resource management and population health trends, it allows for smarter, more efficient policy-making that keeps pace with rapid demographic and epidemiological changes. Ensuring that quantum AI's potential benefits all, especially marginalized communities, remains a vital goal for policymakers and health equity advocates.

While these breakthroughs stand at the threshold of possibility, the journey is just beginning. Continuous research, development, and ethical discourse are crucial to turning the promise of quantum AI into a reality that revolutionizes healthcare. As we look towards the horizon of medical innovation, the promise of quantum AI stands as a beacon—a transformative force that could redefine wellness and healing across the globe, fundamentally improving the quality of life for all.

Personalized Medicine and Quantum Computing

The integration of quantum computing with personalized medicine represents a frontier that could redefine healthcare's landscape. In this transformative era, quantum computing's unparalleled potential to process vast datasets at remarkable speeds can propel precision medicine to new heights. This is not just about enhancing data analysis; it's about understanding intricate biological systems in ways we never thought possible. The complexity of the human genome and the vast array of genetic variations that influence individual health has always posed significant challenges. However, quantum computing may offer the computational power necessary to decode these complexities efficiently.

One of the pivotal benefits of bringing quantum computing into the realm of personalized medicine is its ability to handle enormous volumes of genomic data. Traditional computing struggles with the immense datasets generated by genomics, often taking months to process information required for genome sequencing. With quantum computing, these timelines could potentially shrink to mere days or even hours. This reduction in time is not just a technical improvement; it has profound implications for early diagnosis and personalized treatment plans, where time is often of the essence.

At the core of personalized medicine is the ambition to tailor healthcare solutions to individual genetic profiles. Consider diseases like cancer, where treatment options are increasingly dependent on understanding the genetic mutations of tumors. Quantum computing has the potential to facilitate the rapid comparison of genetic profiles against vast databases of known mutations and potential drug responses. This could lead to highly individualized treatment regimens, maximizing effectiveness while minimizing side effects. Such an approach supports the shift towards medicine that is not just reactive but predictive and preventive.

The intersection of quantum computing and personalized medicine isn't limited to genomics. Proteomics, the study of proteins expressed by a genome, could also benefit immensely. Proteins are the workhorses of cells, and understanding their 3D structures and functions is critical for drug discovery and disease treatment. Classical methods often fall short in predicting protein folding and interactions accurately due to the complexity involved. Here, quantum computing might offer the solutions that have eluded conventional computing, leading to breakthroughs in understanding diseases at a molecular level.

Moreover, pharmacogenomics, which studies how genes affect a person's response to drugs, could be revolutionized by quantum capabilities. The current trial-and-error approach in drug prescriptions could be overtaken by precise, data-driven methods that quantum computing enables. Imagine a future where a patient's complete genetic makeup is swiftly analyzed to predict which medications are most likely to be effective, thus optimizing therapy and reducing adverse reactions. Such advancements can vastly enhance the quality of patient care, shifting the paradigm from broadly generalized medicine to treatments tailored to individual needs.

Beyond improving drug efficacy, quantum computing in personalized medicine holds the promise of significantly advancing our understanding of disease processes. Diseases like Alzheimer's and Parkinson's, with their multifaceted pathologies, could be investigated at a granular level, unveiling previously uncharted pathways and mechanisms. This deeper insight can inform the development of novel therapeutics and interventions, addressing a broader range of conditions that have historically been challenging to treat.

These quantum-driven capabilities in personalized medicine are still emerging, but their potential is undeniable. Collaborations between quantum physicists, bioinformaticians, and medical researchers are crucial to harnessing this technology effectively. Interdisciplinary efforts are already underway, pooling expertise from diverse fields to tackle the inherent challenges and to decipher the substantial volumes of data required to make personalized medicine a reality.

While quantum computing offers exciting possibilities, the journey is not devoid of challenges. Building quantum systems that are stable and scalable enough for healthcare applications is a monumental task. Securing patient data in quantum environments and maintaining privacy and ethical standards are equally critical. As we advance, we must navigate these complexities with foresight and caution, ensuring that the technology benefits everyone equitably.

In conclusion, the marriage of quantum computing and personalized medicine is poised to usher in a new epoch in healthcare. By elevating our capacity to analyze, understand, and utilize biological data at an unprecedented scale, we are moving towards a world where diseases are not just treated but anticipated and prevented. As we stand on the verge of these breakthroughs, there lies a path filled with potential to not only treat illnesses more effectively but to comprehend and improve human health in ways previously unimaginable. Embracing this future holds the promise of personalized healthcare that is as unique as each individual it serves, marking a significant leap towards truly individualized medicine.

AI-driven Diagnosis and Treatment

At the intersection of quantum computing and artificial intelligence stands a monumental shift in the healthcare landscape. The synergy of these technologies is poised to enhance diagnostic accuracy and treatment efficacy in ways previously unimagined, promising not only faster outcomes but also more comprehensive healthcare solutions. The power of quantum AI in the medical realm is particularly noteworthy as it holds the potential to transition from traditional diagnostic models to those driven by advanced computational insights, dramatically altering how medical professionals understand and respond to illnesses.

The promise of an AI-driven approach to diagnosis and treatment begins with data — patient records, imaging results, genome sequences, and more. The current healthcare system generates terabytes of data daily, presenting both a challenge and an opportunity. Traditional methods often lack the capacity to process this information efficiently, which is where the computational prowess of quantum AI comes into play. By exploiting the principles of superposition and entanglement, quantum algorithms can analyze massive datasets at unprecedented speeds, uncovering patterns and correlations that might elude classical computers.

The application of AI-driven diagnosis has set the stage for personalized medicine. Through quantum-enhanced AI, doctors can forecast disease trajectories, identify potential health risks, and recommend tailored preventative measures or treatments based on an individual's unique genetic makeup. Quantum AI models can assimilate vast swathes of genomics data along with lifestyle factors, generating insights that support customizing healthcare down to a molecular level.

This transformation is not just theoretical. In pilot studies, AI systems powered by quantum computing are capable of diagnosing conditions like cancers, cardiovascular diseases, and neurological disorders with remarkable preciseness. These systems can process imaging scans with finer granularity and predictive accuracy, often identifying anomalies that human radiologists might miss. Moreover, the predictive models they create are subsequently refined by a feedback loop from real-world outcomes. This iterative improvement continuously enhances the model's diagnostic capabilities, ensuring more accurate prognoses and treatment suggestions.

Furthermore, the treatment landscape is undergoing a similar evolution. Consider drug discovery, a traditionally time-consuming and costly endeavor. Quantum AI can significantly expedite this process by simulating molecular interactions at a quantum level, predicting how different compounds will affect a biological target. Such simulations could lead to the identification of viable drug candidates in weeks rather than years. This acceleration not only shortens the path to market for new treatments but also increased the variety of treatments available, adapting to the diverse genetic profiles of patients.

Beyond diagnostics and drug discovery, the AI-driven approach also impacts the management and optimization of treatment regimens. AI models can learn from treatment outcomes by integrating patient feedback and sensor data, adjusting protocols with greater precision. This capability is vital in managing chronic conditions where treatments need constant fine-tuning based on the patient's response. Quantum AI can thus enable personalized and adaptive treatment paths, ensuring that each patient receives the most effective care possible.

The advent of AI-driven tools in healthcare also promises to alleviate administrative burdens, thus allowing healthcare providers to focus more on patient care rather than paperwork. Quantum-powered AI stands ready to revolutionize electronic health records, making them more interactive and intuitive while ensuring privacy and security through advanced encryption methods. This transformation leads to a more streamlined healthcare system where information is accessible, secure, and positively utilized.

Despite its potential, the integration of quantum AI into healthcare faces hurdles. Concerns about data security, ethical implications, and equitable access to these advanced technologies must be addressed. It's crucial to establish robust frameworks to guide the deployment and use of AI in a healthcare setting, ensuring that its benefits are not only maximized but also available to diverse populations. This involves careful consideration of privacy laws, data ownership, and the fair distribution of computational resources.

Ultimately, AI-driven diagnosis and treatment, bolstered by quantum computing, heralds a new era in healthcare. It pushes the boundaries of what's possible, offering a glimpse into a world where illnesses are not just managed but proactively prevented, where treatments are not just standardized but personalized. As these technologies mature, they promise to empower medical professionals, enrich patient outcomes, and transform the healthcare experience on a global scale, transcending the limitations of current models to deliver a future where healthcare is as dynamic as the life it serves to protect.

Chapter 8: Quantum AI in Financial Technology

The fusion of quantum computing and artificial intelligence is heralding a transformative era for financial technology. Imagine financial markets operating with unparalleled efficiency, as quantum algorithms process vast datasets to unveil patterns and insights previously hidden to classical computers. This integration enables groundbreaking advancements in financial modeling, where risk assessments become more accurate, and scenarios once too complex to compute are now within reach. Quantum AI's capabilities extend to sophisticated fraud detection systems, swiftly identifying anomalies and safeguarding transactional integrity. This technological leap not only augments financial strategy but also ensures a secure and reliable flow of economic activities. As financial institutions start embracing this hybrid approach, the landscape of finance is poised for innovations that promise increased stability and opportunity, harnessing a potent blend of quantum precision and AI prowess.

Innovations in Financial Modeling

The realm of financial modeling stands on the brink of a technological revolution as Quantum AI converges with financial technology. This marriage promises not only to reshape how models are built and interpreted but also to enhance the predictive power and accuracy of financial forecasts significantly. At the heart of this transformation lies the remarkable ability of quantum computing to handle complex calculations and vast amounts of data with unprecedented speed and efficiency. *Imagine a world where data involving countless variables and stochastic processes are analyzed in mere moments.* This is the landscape Quantum AI is poised to create in financial modeling.

Traditionally, financial modeling has depended heavily on historical data and a series of assumptions to forecast future trends. The complexity grows exponentially when dealing with multifaceted datasets, making traditional computing methods cumbersome and sometimes inadequate. Quantum computing introduces a paradigm shift by its inherent design; qubits allow for superposition and entanglement, enabling a multitude of calculations across different scenarios simultaneously. This offers the potential to open up entirely new perspectives and insights, hitherto inaccessible, to finance professionals.

Consider the implications of this capability in options pricing. Models like the Black-Scholes often underestimate the dynamic nature of volatility in financial markets. Quantum AI can improve these models by enhancing their adaptive capabilities through real-time, iterative re-calibration based on incoming data. This method not only refines the predictive accuracy but also equips traders and financial analysts with actionable insights at speeds never before imagined.

Furthermore, the integration of machine learning algorithms within quantum frameworks adds another layer of sophistication. Quantum Machine Learning (QML) can decipher intricate patterns and correlations in data, transforming raw and complex information into coherent strategies. These enhanced analytical tools can have profound applications in portfolio optimization, where balancing risk and return amid shifting market conditions becomes a highly dynamic and personalized task. The result is a more resilient and adaptable financial ecosystem.

Indeed, with Quantum AI, the granularity of financial models improves dramatically. Financial institutions can harness this power to deliver bespoke solutions to clients. With the capacity to process large, real-time datasets, personalized investment strategies that respond to individual risk appetites and market movements can be swiftly devised and executed. This level of precision would signify a new era in wealth management and personalized finance.

Moreover, the promise of Quantum AI doesn't end with the models themselves but extends to the very way financial data is processed and stored. Quantum computing's potential for unprecedented data compression could revolutionize how financial datasets are archived and accessed, reducing costs and latency in data retrieval. This could lead to a dramatic acceleration in the filter-and-feed mechanisms currently employed by financial analysts worldwide.

The implications for regulatory compliance are equally transformative. The financial sector's stringent regulatory requirements necessitate meticulous reporting and stress testing of financial models. With its capacity to simulate an incredibly vast combination of variables, Quantum AI can automate and enhance these processes—ensuring that regulatory frameworks are not only adhered to but also continually optimized in response to evolving global standards.

In the broader landscape, this innovation could democratize financial advising by reducing disparities in access to cutting-edge financial tools. Institutions traditionally limited by resources and technological expertise could gain foothold through quantum-powered solutions, leveling the playing field and fostering increased competition and innovation in the sector.

Crucially, the impact of these innovations extends beyond mere technical improvements. They herald a philosophical shift in how financial models are perceived, moving from static, historical replicas to living, breathing systems that evolve in real time. This redefines what it means to forecast, transitioning from the realm of uncertain projection to confident simulation—a quantum leap of faith, grounded in profound scientific advancement.

However, this revolution doesn't come without challenges. The transition to Quantum AI imposes the necessity for new skill sets, requiring financial professionals to blend traditional expertise with quantum literacy. As enterprises begin to integrate these technologies, the demand for a new breed of analysts and strategists, equipped with the knowledge of quantum mechanics and AI, will become increasingly pressing. Training the workforce to harness this new technology will be a major focus in the coming years, with substantial collaborative efforts required across financial, academic, and technological sectors.

In conclusion, the innovations quantum computing and AI bring to financial modeling are not just about enhancing computational power; they represent a fundamental shift in our theoretical approaches and practical methodologies. As Quantum AI continues to evolve, it holds the promise to not only rewrite existing financial paradigms but to craft entirely new narratives within the financial tech industry. The future of finance lies not just in the numbers themselves but in the revolutionary tools transforming them into wisdom and opportunity.

Risk Assessment and Quantum Computing

In the realm of financial technology, risk assessment stands as a cornerstone of strategic decision-making. With the advent of quantum computing, the landscape is poised for transformative changes, offering unprecedented capabilities to evaluate and manage risk. Traditional computing methods, faced with limitations in handling complex variables and vast datasets, may soon be overshadowed by the quantum revolution.

Quantum computing introduces an entirely new paradigm through its utilization of qubits—quantum bits—that leverage the principles of superposition and entanglement. This breakthrough allows for the processing of information in ways that classical bits never could. In risk assessment, where a multitude of factors interact in complex, sometimes unpredictable ways, the ability to rapidly explore all potential outcomes in parallel offers a seismic shift in efficiency and accuracy.

The volatility of financial markets demands swift and precise risk computations, tasks at which quantum computers excel. Quantum algorithms, such as those designed for factoring or optimisation, provide quantum computers with the edge they need to tackle problems previously deemed intractable. Credit risk models, for instance, which evaluate the probability of default for borrowers, can be enhanced by quantum techniques that analyze far greater volumes of data than classical models, offering results that are both timely and more accurate.

Consider the task of portfolio optimization. This involves selecting the best investment strategy from a nearly infinite array of possible allocations, subject to various constraints and market conditions. Quantum computing's ability to simulate all these scenarios simultaneously allows for an exploration of solutions on a scale that classical computing can only approximate through time-consuming, sequential processing. As quantum technology matures, we anticipate a revolutionary augmentation of current risk management frameworks.

This immense computational power brings with it both opportunities and concerns. On one hand, the ability to anticipate financial downturns or the impact of market changes could lead to more robust economic systems. On the other hand, the potential misuse of such capabilities, perhaps through manipulation of sensitive financial systems, raises ethical and regulatory challenges that must be navigated carefully.

Moreover, implementing quantum computing in risk assessment isn't akin to flipping a switch. It involves addressing significant technical challenges that include developing suitable algorithms, ensuring system scalability, and mitigating quantum errors. Researchers and industry leaders must work collaboratively to create secure and reliable quantum systems. Investment in robust quantum infrastructure and training a skilled workforce are crucial steps towards the seamless integration of this technology into financial risk processes.

Regulatory bodies will play a critical role in this transition by setting guidelines that ensure the ethical and secure deployment of quantum technologies in finance. Such regulations should aim to strike a balance between fostering innovation and protecting stakeholders from unforeseen risks. Collaborative efforts across the financial industry will be needed to maintain transparency and trust as quantum computing becomes a more prominent force in risk assessment.

Cybersecurity, too, emerges as a pivotal consideration. Quantum systems bring enhanced security risks but also the potential for more powerful cryptographic solutions. Financial institutions must strategically design quantum-safe encryption methods to safeguard sensitive data and transactions from potential quantum threats, especially as quantum computers are predicted to eventually crack classical encryption algorithms.

As we delve deeper into the intricacies of quantum computing's role in financial risk assessment, a broader view reveals its potential impact on societal structures. The capacity to foresee and mitigate systemic risks might lend itself to preventing crises that could ripple across economies, affecting communities at every level. The predictive power of quantum-enhanced models could thus enhance global financial stability.

In essence, the intersection of risk assessment and quantum computing is not merely a technical evolution; it's a paradigm shift that challenges the very fundamentals of how financial risk is perceived and managed. As financial institutions start to tap into this extraordinary power, they are also opening a window to unparalleled advancements—or risks—within the industry. The true potential lies in our commitment to responsibly harness these capabilities, adopting a forward-thinking mindset that's as dedicated to innovation as it is to ethical vigilance.

Ultimately, the journey of integrating quantum computing into risk assessment in financial technology may define the trajectory of financial services moving forward. It's a complex, multi-faceted endeavor requiring both caution and courage, promising innovations that were once the dreams of futurists now coming closer to reality. The stage is set for a financial renaissance powered by quantum insights, urging us to both embrace and prepare adeptly for the changes entailed.

AI for Fraud Detection

In the realm of financial technology, fraud detection has evolved into an intricate dance of cat and mouse, where criminals constantly devise new techniques and technology responds with ever-advancing measures to thwart them. Artificial intelligence (AI), with its ability to analyze vast data sets and uncover patterns, has become a pivotal tool in this ongoing battle. When combined with quantum computing, AI's capabilities grow exponentially, offering powerful solutions tailored to detect fraud before it significantly impacts organizations and individuals alike.

Fraud manifests in numerous forms, from credit card fraud and identity theft to more sophisticated corporate deception. Traditional fraud detection systems rely on predefined rules and patterns, but they often struggle with the adaptability and creativity of fraudsters. AI, empowered by machine learning algorithms, changes the status quo by learning and evolving. It can detect anomalies and outliers, which are often indicators of fraudulent activity, through an ever-improving understanding of what "normal" looks like. This dynamic capability transforms how financial institutions and companies guard against financial crime.

Now, imagine the profound implications when you infuse this advanced AI with the power of quantum computing. Quantum AI possesses the potential to process information at speeds and volumes previously unimaginable. This means that when fraud patterns quickly morph and scale, especially during high transaction periods, quantum-powered AI is not just keeping pace but is often a step ahead. By leveraging quantum bits, or qubits, quantum AI analyzes these transactions in parallel, examining hundreds of variables at once, something classical AI struggles to achieve in real-time.

Consider a large retail company processing thousands of transactions every minute, each a potential target for fraud. Quantum AI can assess each transaction not just for its data accuracy but against multiple dimensions of fraud criteria simultaneously. Attributes such as transaction location, time, amount, and recency of similar transactions are just the beginning. With quantum-enhanced AI, complex datasets become easily navigable terrain, where anomalies are swiftly identified, often intercepted before they manifest into actual fraud events.

Moreover, the predictive analytics capabilities of quantum AI offer a unique advantage. They allow financial entities to anticipate fraud trends before they significantly impact operations. By continuously running simulations and analyzing massive datasets, it predicts potential threats and evaluates the effectiveness of preventive strategies long before malicious efforts unfold. This gives organizations a proactive edge, reducing the need for reactive damage control measures that are often costly both economically and reputationally.

For instance, in the insurance industry, fraud detection is crucial in claims processing. Quantum AI not only checks the veracity of claims against historical data but also models future risks. This predictive power means insurers can offer more competitive rates to honest clients while weeding out fraudulent claims with greater accuracy. Similarly, in e-commerce, where card-not-present transactions are common, quantum AI helps verify each transaction faster and with increased precision, reducing false declines, which can hurt customer trust and company revenue.

Quantum AI's integration into financial infrastructures isn't just a technological leap; it's a transformational shift. It democratizes the detection capabilities, making sophisticated fraud prevention accessible even to smaller organizations that might lack the resources for traditional, high-cost detection systems. This inclusivity is crucial because it secures the broader financial ecosystem and ensures that smaller players are not easy targets for cybercriminals due to limited defenses.

What's particularly fascinating is the synergy between human and machine intelligence in this domain. Quantum AI assists analysts by filtering through noise and presenting the most relevant insights, allowing human experts to make informed decisions swiftly. This collaboration not only increases efficiency but ensures that critical thinking and human judgment are integral parts of the fraud detection process, where nuance and context matter.

Another compelling aspect of AI for fraud detection is its potential for continuous learning. As transactions flow and fraud tactics evolve,

AI systems modify their algorithms, getting smarter without needing constant reprogramming. Quantum-enhanced AI harnesses this learning capability with even greater effectiveness, providing a nearly adaptive shield against new fraud schemes. This adaptability is invaluable in a landscape where threats are perpetually changing.

While the possibilities are substantial, integrating quantum AI into existing systems poses its challenges. Organizations must navigate the complexities of merging two pioneering technologies that require not only technical but regulatory and ethical consideration. Privacy concerns, data security, and compliance with financial regulations are critical areas that necessitate robust strategies to ensure public confidence and compliance. Fortunately, as the technology evolves, solutions for these challenges are being developed concurrently.

In conclusion, AI for fraud detection powered by quantum computing isn't just a concept of science fiction—it's an emerging reality poised to redefine the landscape of financial protection. By amalgamating speed, scale, and intelligence, it provides a formidable framework against the escalating threats posed by fraudsters in the digital age. As we advance, this technological harmony will not only safeguard our financial institutions but will also build a trustful, resilient financial environment essential for future prosperity.

Chapter 9: Transforming Transportation with Quantum AI

In the quest to revolutionize transportation, Quantum AI emerges as a powerful ally, reshaping how we conceptualize mobility and logistics. By leveraging the nuanced capabilities of quantum computing, this cutting-edge synergy offers potent solutions for optimizing traffic systems, dramatically minimizing congestion through real-time data analysis. Beyond mere traffic control, Quantum AI plays a pivotal role in advancing autonomous vehicles, enabling them to navigate complex environments with unprecedented precision and safety. As supply chains grow increasingly intricate, the quantum-enhanced analytical power facilitates efficient logistics and resource distribution, ensuring that goods reach their destinations swiftly and sustainably. In essence, Quantum AI doesn't just promise incremental advancements; it holds the potential to transform the very infrastructure of transport, ushering in an era of unprecedented efficiency and environmental responsibility.

Optimizing Traffic Systems

Our world has become increasingly urbanized, leading to densely populated cities where traffic congestion is a pervasive issue. Traditional methods of traffic control, reliant on fixed algorithms and historical data, often fall short in managing the dynamic nature of urban traffic flow. However, the hybrid power of quantum computing and AI offers novel solutions to transform our traffic systems into smarter, more efficient networks. Through their combined capabilities, traffic can be optimized in ways that were previously unimaginable.

Quantum computing, with its ability to handle complex computations substantially faster than classical computers, is at the heart of this transformation. When dealing with urban traffic, countless variables interact with each other unpredictably. From varying vehicle speeds to differing driver behaviors and pedestrian crossings, the sheer number of potential scenarios is overwhelming. Quantum algorithms, however, can process these variables simultaneously thanks to the principles of superposition and entanglement. This parallel processing capability enables a more robust real-time analysis of traffic data.

Incorporating AI accentuates these advancements by learning and improving from patterns in traffic flow over time. AI systems leverage machine learning techniques to predict congestion points and optimize routes accordingly. When paired with quantum computing's processing power, AI can offer adaptive solutions that respond dynamically to emerging traffic situations, such as sudden road closures or accidents. This synergy can lead to reduced travel times, less fuel consumption, and lower emissions, supporting sustainable urban development.

Consider a citywide network where quantum-powered sensors and AI are integrated at every traffic node. These smart sensors actively collect data on traffic density, weather conditions, and even driver behaviors. Quantum AI systems process this information to identify inefficiencies, dynamically adjusting traffic light patterns and implementing variable speed limits in real-time. Such adaptability ensures a smooth and continuous flow of vehicles, minimizing unnecessary idling or congestion.

This transformation doesn't only pertain to roadways. Public transportation systems stand to benefit substantially as algorithms optimize train schedules and bus routes with precision. For instance, quantum AI can dynamically alter bus routes based on current passenger loads and traffic conditions, ensuring more timely services and efficient energy use. By steering the public towards reliable and efficient mass transit options, overall traffic can be vastly reduced.

Quantum AI's potential in traffic optimization isn't just theoretical; cities worldwide are beginning to usher in pilot programs to test these technologies. Urban centers like Tokyo and Singapore are leading the charge, integrating quantum algorithms to refine traffic management. These small-scale implementations are showing promising results, demonstrating reductions in congestion and improvements in commuter satisfaction.

Yet, bringing quantum AI to traffic systems across the globe involves overcoming significant hurdles. Infrastructure upgrades are needed to support the deployment of this cutting-edge technology. Retrofitting existing traffic control systems with smart sensors and quantum processing units represents a substantial investment. Additionally, a shift towards data-centric governance will be crucial, as the efficacy of these systems depends heavily on the quality and volume of real-time data.

Privacy also becomes a concern in the data-driven management of urban traffic. Handling vast amounts of information about individual driving patterns demands stringent data protection frameworks. Ensuring that privacy protocols keep pace with technological advancements will be essential to maintain public trust.

Looking beyond current challenges, the integration of quantum AI in traffic systems can profoundly impact urban planning and development. City planners can use insights gained from traffic analyses to design roads and public spaces that cater to evolving usage patterns. This data-driven approach will not only streamline existing networks but pave the way for sustainable urban spaces that prioritize the needs of their inhabitants.

Moreover, the implications of optimizing traffic systems extend into economic realms. Improved efficiency in transportation translates to enhanced productivity as the time spent stuck in traffic diminishes. Businesses can better rely on timely logistics, reducing costs associated with delays and enabling faster delivery services. This technological leap could thus spur economic growth and elevate living standards.

The journey towards advanced traffic optimization with quantum AI is a testament to human ingenuity and our relentless pursuit of innovation. As these technologies mature, they promise to reshape how we navigate our environments, turning the cluttered byways of today into the seamless mobility arteries of tomorrow.

With continued research and development, the vision of harmonious, efficient, and sustainable transportation systems isn't far from realization. As urban landscapes evolve, grounded in the principles of quantum and AI, they serve as a reflection of the potentials unlocked when groundbreaking technology is harnessed with a conscientious vision for the future.

Development of Autonomous Vehicles

The exploration of autonomous vehicles (AV) represents a blend of technological innovation and revolutionary change, poised to reshape the landscape of transportation. As we stand at the precipice of the next era in automobility, it's clear that Quantum AI is set to play a pivotal role in its development and transformation. The fusion of quantum computing and artificial intelligence delivers unprecedented computational powers and innovative algorithms that enable machines, notably vehicles, to perceive and navigate the world with an intelligence once thought to be the realm of science fiction.

Autonomous vehicles rely heavily on a multitude of sophisticated technologies working harmoniously. LIDAR, radar, cameras, and other sensor technologies generate a deluge of data that demands real-time processing and decision-making. Quantum AI enters the scene as a transformative force, enhancing the ability to process complex datasets and execute concurrent computations at exponentially higher speeds than classical systems. This leap in capability heralds a new paradigm in how AV systems are designed and operated.

Central to the quantum advantage in AV technology is the concept of superposition and entanglement, two fundamentals of quantum mechanics. These principles enable quantum computers to hold and process multiple possibilities simultaneously, providing autonomous systems with the ability to evaluate a vast array of environmental factors with incredible efficiency. When driving conditions change suddenly or unpredictably, quantum-enhanced algorithms can swiftly adjust and react, prioritizing safety and minimizing risk.

The challenge of translating vast sensor data into actionable intelligence garners a major boost from quantum AI. Traditional AI models, while robust, often struggle with the complexities involved in real-time traffic environments. The quantum-enhanced neural networks provide richer, more nuanced interpretations of sensor data, allowing AVs to better understand their surroundings, predict other vehicle movements, and make precise navigation decisions.

One can't overlook the profound implications that these advancements hold for urban mobility and logistics. Autonomous vehicles fueled by quantum AI present unprecedented opportunities in streamlining traffic, reducing congestion, and optimizing route planning. The promise of a fleet of AVs operating seamlessly within a city infrastructure could drastically lower emissions and energy consumption by optimizing travel routes and reducing idle times.

Moreover, as global populations continue to urbanize, the integration of intelligent AV systems entwined with smart city designs could lead to more sustainable urban development. Cities powered by AI-driven transportation networks might evolve into dynamic entities, where traffic lights adapt to real-time conditions, parking is autonomously allocated, and public transportation systems are efficiently synchronized.

However, it's not just urban transportation that's poised for disruption. Autonomous freight and delivery services stand to gain as well. Quantum AI can markedly enhance the logistics industry's efficiency, with AV trucks optimized for energy consumption, safety, and timely deliveries. This development could signify a substantial reduction in operational costs, which, in turn, might decrease prices for consumers and increase profit margins for businesses.

Safety is a fundamental pillar in the development of autonomous vehicles. Here, quantum AI plays an instrumental role by significantly reducing accident risks. Through enhanced data processing and rapid pattern recognition, quantum-augmented AV systems predict and counteract potential hazards more accurately. For instance, AVs could anticipate an imminent collision based on subtle environmental cues and take evasive actions faster than human reflexes.

The integration of quantum AI in AVs does present substantial challenges, particularly in terms of data privacy and security. The interconnected nature of these vehicles, handling significant volumes of sensitive data, could attract cyber threats. Ensuring robust cybersecurity protocols is crucial to safeguard against such vulnerabilities. Quantum cryptography may offer solutions, providing a formidable defense against potential breaches.

Beyond logistics and safety, there's an often-discussed societal transformation that autonomous vehicles might bring. The technology has the potential to democratize transportation, granting mobility to those who were previously disadvantaged, such as the elderly and disabled. Quantum AI can ensure that the promise of efficient, equitable access to transportation becomes a global reality.

The journey from prototype to widespread adoption of AVs will not be without obstacles. Regulatory landscapes need to be navigated deftly, with policies adapted to embrace technological innovations without compromising public accountability. Standards for safety, ethics, and interoperability across regions must be harmonized to facilitate the global deployment of quantum-powered AVs.

The vision of a fully autonomous future is within reach, and quantum AI's role in this transformation cannot be overstated. By bridging quantum computing's vast potential with AI's adaptive capabilities, we're unlocking new realms of possibility, setting the stage for a revolution in how transportation is perceived and utilized.

Envision a future where cars communicate in seamless symbiosis—predictive, efficient, almost alive to their surroundings. It's a vision where roads are safer, journeys are smoother, and our carbon footprint is minimized. As we continue to tread this exhilarating path, the eventual harmonization of quantum AI with autonomous vehicles will likely propel us into a smarter, more connected world where the possibilities for innovation and societal benefit are boundless.

Efficient Supply Chain Solutions

In a world that's increasingly interconnected, supply chains form the backbone of global commerce. As we move into an era dominated by quantum AI, the potential for transforming these vital networks becomes both apparent and profound. With businesses under pressure to deliver efficiently and sustainably, harnessing the unique capabilities of quantum computing in tandem with AI can reshape supply chain management in remarkable ways.

The traditional supply chain is fraught with complexity. Managing inventory levels, forecasting demand, optimizing logistics, and minimizing delays requires sophisticated strategies and tools. Enter quantum AI: a transformative force capable of revolutionizing each of these processes. By leveraging the principles of quantum mechanics, such as superposition and entanglement, quantum AI systems can process vast datasets at unprecedented speeds. This allows for the creation of models that predict demand more accurately, streamline logistics, and manage resources with unparalleled efficiency.

Quantum AI can substantially enhance the efficiency of routing and logistics within supply chains. For instance, the complex problem of optimizing delivery routes across multiple destinations can often baffle classical systems. Known as the "travelling salesman problem," it involves finding the shortest possible route that visits a list of locations and returns to the origin point. Classical algorithms struggle with such problems when the number of destinations increases. Quantum algorithms, however, excel by swiftly testing a multitude of routes in parallel, an approach grounded in quantum superposition. This capability not only saves time but also reduces fuel consumption and associated carbon emissions, significantly boosting the sustainability of logistics networks.

Another critical aspect lies in inventory management. Traditional approaches attempt to balance supply with fluctuating demand, yet forecasts aren't always reliable. Quantum AI can model complex systems dynamically, considering a myriad of variables that influence market demand. By analyzing these variables concurrently, businesses gain a deeper understanding of market trends, leading to improved decision-making about stock levels and procurement strategies. By ensuring that the right products are available when needed, companies can successfully avert costly overstock or stockouts.

Supplier risk management also sees benefits from quantum-enhanced analytics. Suppliers are interconnected and interdependent, forming intricate networks with vulnerabilities that classical AI might not detect promptly. With its capability to assess risk from various angles simultaneously, quantum AI can anticipate disruptions, such as geopolitical issues or natural disasters, providing early warnings and preventive strategies. This proactive risk management fosters more resilient supply chains, reducing potential downtime that might have previously stalled production lines or distribution.

Aspects of machine learning integrated with quantum computing further elevate supply chain strategies. Quantum-enhanced neural networks offer enhanced predictive analytics, far outperforming classical counterparts. This advancement not only affects logistics and inventory but also improves overall quality control throughout production processes. By identifying and correcting inefficiencies and anomalies in real-time, quantum AI ensures that supply chains operate at peak efficiency.

Another breakthrough is seen in the optimization of warehouse operations. The optimal placement of items and streamlining of picking processes are significant challenges. Quantum AI can simulate numerous configurations rapidly, identifying solutions that dramatically reduce the time and costs associated with order fulfillment. By enabling faster, more reliable operations, supply chains become agile and responsive to market changes, leading to increased customer satisfaction.

One fascinating potential application is in devising new strategies for supply chain sustainability. Quantum AI can simulate the impact of different logistics methodologies on carbon footprints, promoting eco-friendly choices. As businesses face growing pressure to adopt greener practices, these quantum-driven insights enable more strategic moves toward sustainability without sacrificing efficiency or profitability.

While the promise of quantum AI in supply chains is immense, its practical implementation still requires overcoming technical and economic barriers. Building robust quantum systems that can reliably manage real-world tasks remains a significant challenge. However, ongoing research and collaboration among technologists, businesses, and policymakers pave the way for overcoming these hurdles. As these quantum systems become more feasible, their integration into existing digital infrastructures will follow, transforming blueprints from theoretical explorations into practical solutions.

In conclusion, quantum AI holds the key to reimagining supply chain management. Its ability to process complex data efficiently offers unparalleled improvements in logistics, inventory, supplier management, and sustainability. The journey of integrating these capabilities into the fabric of global commerce represents not just progress, but the dawn of a transformative era in business operations. With continued advancements, the once elusive goal of a fully optimized, sustainable, and resilient supply chain network could soon become reality, driven by the unprecedented potential of quantum AI.

Chapter 10: Environmental Impact and Solutions

In an era where the planet's health is more critical than ever, the fusion of quantum computing and AI offers transformative solutions to environmental challenges. These advanced technologies are spearheading innovations in climate modeling, enabling unprecedented precision in predicting weather patterns and the effects of global warming. With quantum algorithms, we can process vast data sets at an unparalleled speed, aiding in the development of sustainable energy solutions that could reshape our reliance on fossil fuels. Moreover, the potential of AI to enhance biodiversity conservation through monitoring and predictive analytics is vast, providing vital insights into species protection and ecosystem management. As we harness the synergy of quantum and AI technology, we're not just envisioning a sustainable future—we're creating the tools to achieve it.

Quantum AI for Climate Modeling

Quantum AI stands at the forefront of revolutionizing how we tackle one of humanity's greatest challenges: climate change. By merging the computational prowess of quantum computing with the adaptive intelligence of AI, we're seeing a transformative approach to environmental modeling. This isn't just about crunching numbers faster; it's about uncovering patterns and solutions previously hidden in the chaos of climate data.

Understanding climate dynamics involves enormous datasets, from atmospheric to oceanic models, and everything in between. Traditional supercomputers strain under such complexities, struggling with limitations in data processing and memory. Quantum computing offers an antidote to these limitations. With quantum bits, or qubits, we're not confined to binary restrictions, allowing for simultaneous processing of multiple possibilities. This quantum advantage means vast improvements in how we can simulate and predict climate patterns.

Artificial Intelligence comes into play by enhancing our ability to interpret and refine these predictions. Machine learning algorithms, particularly those enhanced by quantum processes, can analyze these qubit computations to predict outcomes more accurately. Imagine AI models that learn at an exponential rate, constantly improving their predictive capabilities by discovering emerging patterns in climate data that humans alone might miss.

One way Quantum AI is making headway is by improving weather forecasts. Accurate forecasts are crucial not only for daily life but for preparing for extreme weather events—which are increasing in frequency and severity due to climate change. By refining models with real-time data, Quantum AI could potentially enhance lead times for natural disasters, giving communities more time to prepare and react, ultimately saving lives.

Furthermore, Quantum AI can help with long-term climate modeling. Climate forecasts now deal with high degrees of uncertainty, often because our models are simplified to make them computationally feasible. But with quantum-enhanced AI, we can run intricate simulations that account for a greater number of variables and interactions. This could lead to more reliable predictions about how different regions will be affected by shifting climates, informing better policy decisions.

Another valuable application is in optimizing energy usage and reducing carbon footprints. Quantum AI can provide insights into complex systems like power grids, identifying inefficiencies, and suggesting ways to shift energy consumption towards more sustainable practices. This has a direct impact on reducing global emissions, aligning with international goals like those set by the Paris Agreement.

Of course, integrating Quantum AI into climate research doesn't come without challenges. Quantum computers are still in their infancy, with limited qubits and coherence times that make them difficult to scale. Moreover, developing Quantum AI algorithms is a daunting task that requires cross-disciplinary expertise. However, the potential impact on climate modeling justifies the effort and investment.

The collaborative nature of this endeavor cannot be understated. Effective Climate Modeling using Quantum AI will rely on partnerships between governments, academia, and industry. By combining resources and expertise across sectors, we can accelerate breakthroughs and push beyond the current boundaries of climate science.

Quantum AI also invites a broader ethical conversation around transparency and data privacy. As we explore these new territories of technology, ethical guidelines need to ensure that environmental data used in these quantum models is gathered and applied responsibly. The resulting insights should be accessible, allowing for informed public discourse and policy-making.

In conclusion, Quantum AI for Climate Modeling holds the potential to renovate our approach to understanding and combating climate change. As our computational tools evolve, so too should our strategies for addressing the most pressing environmental issues of our time. Through the synergy of quantum computing and AI, we're inching closer to a future where climate predictions are not just hopeful estimates, but accurate, actionable insights that pave the way to a sustainable future.

Sustainable Energy Solutions

As we continue to grapple with climate change, the quest for sustainable energy solutions has never been more urgent. The intersection of quantum computing and artificial intelligence (AI) offers promising pathways to reshape our energy paradigms, highlighting innovations that could significantly reduce humanity's environmental footprint. Given the enormous potential these technologies hold, it's imperative to explore how quantum AI can help in harnessing renewable energy sources more efficiently, optimizing energy distribution, and predicting energy needs.

Firstly, one of the most significant contributions lies in renewable energy sector optimization. Quantum computers, with their extraordinary processing capabilities, can model complex systems more accurately and manage vast datasets effectively. This capability is crucial in refining the efficiency of solar and wind energy conversions. By simulating various energy scenarios, quantum algorithms can predict and optimize the placement of solar panels or wind turbines, resulting in maximum energy generation with minimal environmental disruption. The AI-integration would further enhance this process by enabling swift adaptations to changing environmental conditions, ensuring a steady and reliable energy supply.

Moreover, AI-driven insights can facilitate better energy management systems. As cities grow and demand for energy surges, managing distribution efficiently becomes a critical challenge. Quantum-enhanced AI algorithms offer solutions by optimizing energy grid operations, balancing load demand, and minimizing transmission losses. This proactive management ensures that energy is conserved and distributed intelligently, leading to reduced carbon emissions and enhanced reliability.

Another crucial area is the acceleration of breakthroughs in energy storage technologies. Effective energy storage remains a major bottleneck in the wider adoption of renewable energy sources. Here, quantum computing's potential to simulate molecular structures and chemical reactions may revolutionize the development of advanced batteries. AI can analyze these quantum simulations to identify materials and structures for better and more efficient energy storage systems at a fraction of the cost and time previously required.

Taking a broader view, quantum AI opens doors in powering innovative solutions for carbon capture and utilization. Quantum computing may reveal new catalysts and processes for converting captured carbon dioxide into useful chemicals or fuels. AI could further help in identifying optimal conditions and processes for these conversions, offering a dual benefit—reducing greenhouse gas emissions and creating value from waste products.

Understanding energy needs and predicting future demands are areas where quantum AI can deliver transformative impacts. Energy forecasting—critical to ensuring sustainable energy supply—can be significantly improved using quantum-enhanced AI models. These models process large-scale data to predict trends with unmatched precision, making it possible to anticipate and act on future energy needs while minimizing unnecessary energy production and resource wastage.

Additionally, the integration of quantum AI could transform smart grid systems, making them more adaptive and resilient. Current grid systems are limited in their ability to integrate disparate renewable energy inputs. Quantum computing, coupled with AI, can enable real-time analysis and decision-making, ensuring that renewable sources are utilized to their fullest potential while maintaining grid stability.

Furthermore, AI's pattern-recognition capabilities, when augmented with quantum techniques, provide advanced diagnostics opportunities. This enables the early detection of potential issues within power plants or grid lines, ensuring timely maintenance and preventing energy disruptions. The reliability and efficiency garnered via predictive maintenance could dramatically improve operational performance and sustainability metrics.

The role of policy cannot be understated in guiding the effective deployment of these technologies. Quantum AI's potential can only be realized with coherent strategies that facilitate innovation while safeguarding ethical considerations. Policymakers need to develop frameworks that address data privacy, transparency, and security—all of which are paramount in the age of intelligent computing.

Collaboration across sectors and disciplines is equally essential to bring these energy solutions to life. Academic institutions, governments, and private sectors should unite to forge new alliances, fuel innovation, and build infrastructures that support the broad implementation of quantum AI-driven sustainable solutions.

Quantum AI stands at the forefront of making sustainable energy a feasible and pragmatic reality. While we are still in the early stages, the fusion of these groundbreaking technologies heralds an era where efficiency and sustainability are not mutually exclusive but harmoniously intertwined. Exploring the myriad avenues quantum AI offers for energy conservation and sustainable growth could signal a pivotal leap toward a cleaner and more resilient future.

Protecting Biodiversity with Technology

In our quest to harness technology to safeguard the planet's biodiversity, both quantum computing and artificial intelligence (AI) emerge as pivotal allies. With ecosystems around the world under threat from human activity, climate change, and other inducing factors, there's an urgent need to leverage cutting-edge technology to devise sustainable solutions. This section explores how these advanced technologies can play a crucial role in biodiversity conservation, driving solutions that go beyond traditional methods.

AI, with its immense data-processing capabilities, can be utilized to monitor ecosystems and species with unprecedented precision. For instance, machine learning algorithms can analyze satellite images to track deforestation or habitat changes with pinpoint accuracy. This real-time monitoring enables conservationists to respond faster to threats, making proactive interventions possible. The integration of AI in camera traps and drones allows for more efficient wildlife surveys, enabling the identification and tracking of species across challenging terrains, often in places where humans can't easily venture.

Moreover, quantum computing can transform the landscape of ecological modeling. Traditional computing has its limitations in processing the complex, interdependent variables within ecosystems. However, quantum computers, with their ability to evaluate numerous possibilities simultaneously, can provide highly accurate simulations of intricate ecological models. This capability can assist in predicting how changes in one part of an ecosystem might ripple through the rest, thereby helping conservationists make informed decisions.

Another promising frontier is the use of technology in genetic conservation. AI algorithms can analyze vast genetic databases to identify species that are at risk of genetic bottlenecking. Quantum computing can further enhance these processes by rapidly running simulations or analyzing genetic data to help understand the potential future of species under various environmental scenarios. It's this synergy of AI and quantum computing that holds the key to revolutionizing conservation genetics, potentially discovering strategies to bolster the resilience of vulnerable species.

The importance of biodiversity extends beyond conservation circles, reaching into global agricultural, pharmaceutical, and industrial sectors, where the discovery of novel organisms or genetic traits can translate into breakthroughs in crop resistance, drug development, and sustainable materials. Here, quantum computing and AI can accelerate research and discovery. By sifting through vast biological data quicker than ever before, these technologies might uncover new insights that improve ecosystem services and contribute to biodiversity's intrinsic and practical value.

While the potential of these technologies is enormous, their implementation is not devoid of challenges. Ethical considerations must be addressed, particularly concerning data privacy and the potential for algorithmic bias, which could overlook or misinterpret critical biodiversity data. As with many technological advancements, there's a learning curve that requires interdisciplinary collaboration among ecologists, technologists, and policymakers to ensure ethical standards guide biodiversity tech initiatives.

Additionally, there's a need for building robust infrastructures capable of handling the computational demands of these advanced technologies. Access to quantum technology is still limited, and as it becomes more widespread, establishing partnerships between tech companies, research institutions, and governments will be essential to democratize access and drive meaningful impact on a global scale.

Emphasizing education and awareness is equally important. By involving local communities in the deployment and use of these technologies, we ensure that conservation efforts are rooted in local knowledge and needs, fostering sustainability and greater acceptance. These technologies also hold the potential to inspire and educate future generations, encouraging a new generation of tech-savvy conservationists who are as comfortable with drones and data analytics as they are in nature.

As we explore this new frontier, it's essential to remain vigilant about potential drawbacks and continuously refine our approaches. Studies should focus on long-term impacts and sustainability of tech-driven conservation projects. While technology provides us with sophisticated tools, it is our responsibility to wield them wisely in the ongoing battle to preserve the rich tapestry of life on our planet.

In conclusion, the intersection of biodiversity protection with cutting-edge technology like AI and quantum computing offers a beacon of hope. These innovations provide us with a powerful arsenal to counter the biodiversity crisis by empowering conservation efforts around the globe. As we stride into this integrated future, embracing these technologies promises to unlock new horizons in the enduring commitment to save our world's natural heritage.

Chapter 11: Quantum AI in the Manufacturing Industry

Quantum AI stands poised to redefine the manufacturing industry by injecting unprecedented efficiencies into production processes. Through the synergetic application of quantum computing and AI, factories can significantly enhance their production throughput and precision. The complex dynamics of inventory management, which have traditionally been a logistical nightmare, can be streamlined with quantum algorithms that offer optimal solutions faster than conventional systems ever could. Moreover, AI-driven quality control systems that once relied on traditional computational models now tap into quantum-enhanced learning capabilities, ensuring that products meet the highest standards while minimizing waste. This seamless integration of Quantum AI into manufacturing does not just represent an incremental improvement; it heralds a transformative leap, enabling manufacturers to meet the ever-growing demand for customization and quality while reducing time to market. As companies embrace these quantum-driven tools, they will find themselves at the leading edge of a new industrial revolution, empowered to craft more sustainable, intelligent, and adaptive production ecosystems.

Enhancing Production Processes

In the bustling landscape of the manufacturing industry, the integration of Quantum AI promises to unshackle the constraints of conventional production processes. This transformative fusion of quantum computing and artificial intelligence isn't just about making things faster. It's an opportunity to perform tasks in ways we've never considered, enabling manufacturers to leapfrog traditional limitations and design processes that are fundamentally smarter and more efficient.

For starters, the capability of Quantum AI to handle immense datasets with remarkable speed and accuracy opens the door to highly optimized production schedules. Traditional production planning relies heavily on approximations and linear models due to computational limitations. But with quantum computing's inherent ability to process complex variables in superposition, manufacturing can achieve scheduling efficiencies that were previously deemed impossible. When these quantum computations are then refined and guided by artificial intelligence, the results are both efficient and astoundingly precise.

Imagine a factory capable of dynamically adjusting its operations in real-time, responding seamlessly to fluctuations in demand, resource availability, or supply chain disruptions. This is not science fiction. Quantum AI, with its capacity to simulate vast, interconnected systems rapidly, can provide predictive insights, allowing managers to reconfigure their operational strategies on the fly. Such capabilities ensure minimal downtime and maximum throughput, vital metrics in this competitive industry.

But it's not just the speed of decision-making that Quantum AI revolutionizes; it's also the quality of those decisions. Traditional process optimization often struggles with finding the perfect balance between conflicting objectives—cost minimization, quality assurance, and environmental sustainability, to name a few. Quantum AI can model these competing demands concurrently, offering solutions that aren't just efficient but also holistic. For instance, advanced simulations can help identify energy-efficient production methods without compromising on speed or quality, aligning with the pressing need for ecological conservation in manufacturing.

Consider the machine learning paradigms that drive AI advancements—neural networks and deep learning models that mimic human decision thinking. In a quantum context, these models gain a significant edge. Quantum-enhanced machine learning can not only identify potential bottlenecks and inefficiencies but can also propose innovative solutions that go beyond the traditional constraints of linear problem-solving. This allows for the possibility of self-improving processes where the entire production line becomes an adaptive, intelligent ecosystem.

Furthermore, the integration of Quantum AI into production doesn't necessitate an overhaul of existing systems. Hybrid models that incorporate both classical and quantum processes allow for a gradual transition, where conventional systems benefit from the overarching guidance of quantum-empowered AI. This interoperability ensures that industries at various stages of technological adoption can harness the benefits without massive disruption.

One fascinating application of Quantum AI is in the realm of materials science, vital to production processes. The ability to discover or design new materials rests heavily on the understanding of molecular interactions—an area where quantum computing excels. By solving complex equations that describe these interactions, Quantum AI can predict new material behaviors, leading to the invention of substances that are lighter, stronger, and more adaptable than anything previously available. This could revolutionize industries like aerospace, automotive, and electronics, where material properties are paramount.

Meanwhile, quality assurance, another cornerstone of manufacturing, can be dramatically enhanced with Quantum AI. By implementing continuous real-time analysis, manufacturers can maintain exceptionally high standards without the typical time delays associated with quality checks. Quantum AI's ability to recognize patterns in complex data sets can lead to rapid identification of defects or inefficiencies, reducing waste and ensuring consistently high-quality outputs.

Ultimately, the vision for Quantum AI in enhancing production processes is one that not only transforms discrete elements of manufacturing but also reimagines the entire system. In doing so, it empowers manufacturers with tools to act with unprecedented agility and foresight, all while fostering innovation at every level of operation. This paradigm shift holds the potential to push the boundaries of what manufacturing can achieve, opening doors to limitless possibilities previously confined to theoretical discussions.

The journey towards fully integrating Quantum AI in manufacturing is not without its challenges and will require strategic planning, investment, and a willingness to embrace change. However, by taking bold steps forward, the manufacturing industry can position itself at the forefront of the technological revolution, ensuring continued growth and sustainability in an ever-evolving global landscape.

AI-driven Quality Control

In the vibrant landscape of manufacturing, maintaining high-quality standards is more than just a goal—it's a necessity. With the integration of *Quantum AI* into quality control processes, the industry stands on the brink of unprecedented innovation. Unlike traditional approaches, AI-driven quality control leverages advanced algorithms to enhance precision, efficiency, and adaptability. It's an evolution that not only improves the production line but reshapes how manufacturers assess, manage, and ensure quality.

Quality control in manufacturing has often relied on human inspection and predefined statistical methods, both of which have limitations in speed and accuracy. Here, AI shines with its ability to process enormous troves of data quickly, draw insights, and learn from patterns—all in real-time. Quantum computing supercharges these AI capabilities by managing data with astonishing speed and accuracy, turning once-insurmountable problems into solvable challenges.

One primary advantage of AI-driven quality control is its ability to detect anomalies at a micro level. Using sophisticated image recognition software paired with quantum-enhanced analytics, AI systems can identify defects invisible to the human eye or traditional sensors. This technology doesn't just find faults; it anticipates them. By analyzing historical data and current production metrics, AI systems predict potential problem areas before they manifest, allowing for proactive measures rather than reactive fixes.

Consider a scenario in a semiconductor plant where every microchip must meet exact specifications to function correctly. Conventional quality checks could overlook minuscule defects. However, AI-driven systems, equipped with quantum computing power, can examine each microchip at a granular level, ensuring defects are spotted and rectified before the chips move to the next stage of production. This reduces waste, increases yield, and minimizes costly recalls.

Moreover, AI-driven quality control can adapt and evolve with the manufacturing line. Learning from each cycle, AI adapts its criteria and thresholds for quality assurance, refining its models with each iteration. This dynamic approach contrasts sharply with static, rule-based systems, which don't evolve with changing production environments or new product introductions.

Integrating AI into quality control also facilitates a more nuanced approach to feedback and continuous improvement. Rather than merely sorting products into pass/fail categories, AI systems provide comprehensive data that manufacturers use to refine processes and enhance product quality. These insights drive innovations in design, materials, and production techniques, ultimately leading to better products and more efficient manufacturing processes.

In addition, the use of AI and quantum computing can enhance the scalability of quality control operations. Large-scale manufacturing facilities can implement uniform quality standards across multiple sites, ensuring consistent product quality regardless of location. This capability is particularly beneficial for global operations that demand stringent adherence to local regulations and consumer expectations.

The global manufacturing industry's logistical complexities also find relief through AI-driven solutions. By reducing the margin for error, companies not only ensure compliance with industry standards but also build trust with consumers and stakeholders. The high levels of accuracy and reliability afforded by AI in quality control reinforce brand reputation and consumer loyalty.

Furthermore, sustainability emerges as a significant benefit in AI's role within quality control. By minimizing waste through precise defect detection and improved process efficiencies, manufacturers can contribute to environmental goals. Reduced material wastage translates into lower resource consumption, aligning with both corporate responsibility and global sustainability objectives.

Challenges remain, of course. Implementing AI-driven quality control systems requires substantial investment in technology and training. There are also questions regarding data security, given the vast amounts of sensitive information these systems handle. However, the potential gains far outweigh the hurdles, rendering AI-driven quality control not just a competitive advantage, but a fundamental component of future manufacturing strategies.

As the manufacturing industry marches towards a future where quantum AI plays a pivotal role, one thing becomes clear: quality control will not just be about meeting standards but continuously redefining them. The combination of AI and quantum computing paves the way for a new era of precision manufacturing, where defects are a rarity and excellence is the norm. Embracing these technologies will not only improve product quality but also redefine the benchmarks of what's possible in manufacturing.

Quantum Solutions for Inventory Management

In the fast-paced world of manufacturing, managing inventory efficiently can often make or break a business. Inventory management is a complex task that involves ensuring the right products are available at the right time and in the right quantities. Traditional methods—reliant on classical computing and algorithmic approaches—can only take us so far. Enter quantum computing, a technology promising to revolutionize the field with its unprecedented computational power and advanced problem-solving capabilities.

Let's delve into why quantum computing stands out. Conventional inventory systems are predominantly driven by algorithmic operations that struggle with optimizing tasks. These tasks, which demand intricate simulations and predictions, are naturally suited for quantum computers. Quantum computing, with its ability to perform calculations at unimaginable speeds, can examine a virtually infinite combination of variables simultaneously. This enables it to generate solutions to complex problems that would leave classical systems struggling.

The allure of quantum computing isn't just in speed but also in the sheer accuracy and depth of analysis it offers. By using principles like superposition and entanglement, quantum solutions can consider every possible state of a variable at once. This transforms how we might approach predictive analytics in inventory management. The ability to model various scenarios with high accuracy means manufacturers can predict demand fluctuations more precisely, minimizing waste and reducing storage costs.

Moving beyond mere predictive analytics, quantum computing opens new doors in the realm of inventory optimization. Optimization in this context is about finding the best possible path for inventory flow that minimizes costs and energy usage while maximizing efficiency and responsiveness. For instance, imagine a manufacturer adjusting production schedules and supply chain logistics in real time based on ongoing insights drawn from quantum-powered algorithms. Such dynamic adaptability is currently a distant dream for conventional systems.

Moreover, quantum algorithms can enhance inventory traceability, a critical component in globalized supply chains where components and products travel across multiple geographies. As global supply chains expand, ensuring every part of the chain is visible and accounted for becomes increasingly challenging. Quantum computing can manage this task with ease, enhancing transparency and accountability across the entire supply network.

One of the most exciting prospects of integrating quantum solutions into inventory management is the ability to effectively manage uncertainties. With quantum technologies, manufacturers can simulate various risk factors and inventory scenarios, allowing them to develop proactive strategies rather than purely reactive ones. This transforms inventory management from a largely heuristic, experience-based practice to a more predictive and data-driven science.

For example, a manufacturer could use quantum computing to simulate the effects of potential supply chain disruptions. Quantum models could predict which suppliers might be affected by environmental factors, geopolitical instability, or market volatility. By accurately forecasting these disruptions, businesses can pivot quickly, sourcing alternatives or rerouting logistics to mitigate impact, ensuring continuity and resilience even in the face of uncertainties.

Implementing quantum solutions does come with its challenges. The nascent technology still faces hurdles in terms of hardware development and error correction. Ensuring that quantum computers can be effectively and affordably scaled for broad commercial use in industries like manufacturing is an ongoing journey. Furthermore, the integration process involves not just technical challenges but also reshaping business workflows and retraining personnel, which can be daunting but ultimately rewarding.

Despite these obstacles, forward-thinking companies are already investing in these technologies, establishing collaborations with quantum tech firms, and developing pilot programs to test and perfect quantum-integrated systems. They understand that tapping into the power of quantum isn't just about staying ahead — it's about redefining what inventory management can accomplish.

As more advances occur in this field, quantum solutions are set to redefine data management central to inventory processes. The combination of AI capabilities with quantum computing introduces a level of intelligence that neither technology alone could achieve. AI can help interpret the vast outputs generated by quantum computations, providing actionable insights to decision-makers.

This synergy allows for modeling scenarios such as demand spikes due to sudden market trends or optimizing the distribution network in real time to minimize transport delays or costs. As a result, companies could drastically reduce overstock and stockouts, maintaining a perfect balance in inventory levels that translates to significant cost savings and enhanced customer satisfaction.

Looking ahead, we anticipate a future where the fusion of quantum computing and AI becomes integral to every facet of inventory management, laying the groundwork for a new era of intelligent, adaptive, and deeply integrated manufacturing processes. Businesses that embrace this transformation can expect to unlock unprecedented efficiencies, resilience, and agility in their operations — a compelling proposition for any leader eager to pioneer in the digital age.

While we stand on the cusp of these transformative technologies, the path forward requires bold vision and collaborative effort. By uniting interdisciplinary expertise, investing in research and development, and remaining curious about the unfolding potential of quantum solutions, the manufacturing industry can not only keep pace with innovation but lead the charge toward a smarter, more sustainable future.

Chapter 12: Advances in Natural Language Processing

In recent years, natural language processing (NLP) has witnessed rapid advancements, fueled by the convergence of traditional AI and the nascent field of quantum computing. Seamlessly blending these technologies, NLP now teeters on the edge of monumental breakthroughs in understanding and generating human language with unprecedented accuracy. Traditional neural networks have made substantial progress, yet quantum computing introduces an entirely new dimension of computational power and efficiency, offering fresh avenues to tackle complex linguistic models and vast datasets. This emerging synergy holds the promise of transforming tasks such as language translation and interpretation, making them more intuitive and culturally nuanced than ever before. With quantum-enhanced systems, the future prospects of NLP technologies are not just incremental—they're revolutionary. Professionals and enthusiasts alike are called to imagine a world where language barriers dissolve into mere challenges, inviting collaboration and creativity across cultures. This chapter situates readers at the threshold of a linguistic evolution, one that reshapes how machines perceive, process, and understand human interaction, heralding a new era of communication possibilities.

Quantum Computing for NLP

Natural Language Processing (NLP) stands as one of the most transformative technologies in the modern era, enabling computers to understand, interpret, and generate human language in a way that is meaningful and contextually relevant. The incorporation of quantum computing into NLP holds the promise of redefining this landscape altogether. By harnessing the power of quantum mechanics, we can tackle some of the most challenging problems in NLP, potentially surpassing the capabilities of classical computational approaches.

At its core, quantum computing leverages the principles of superposition and entanglement, allowing quantum computers to process information exponentially faster than classical computers. This unique capability opens up new avenues for solving problems that require massive computational power, such as understanding nuanced human language. In traditional NLP models, tasks like language translation, sentiment analysis, and semantic understanding become computationally intensive as the datasets grow larger. Quantum algorithms could provide the efficiency needed to process complex linguistic patterns at an unprecedented scale.

One of the most compelling applications of quantum computing in NLP is in enhancing the performance of machine learning models. Quantum-enhanced machine learning can lead to profound improvements in training times and model accuracy. The ability of quantum computers to explore vast solution spaces simultaneously offers a new paradigm where models that traditionally require enormous datasets and prolonged training could see notable optimization in both time and resources.

However, it's not all straightforward. Implementing quantum computing in NLP has its challenges. Current quantum hardware is still in an embryonic stage, known as the Noisy Intermediate-Scale Quantum (NISQ) era. NISQ devices, while powerful, suffer from noise and decoherence, presenting hurdles in reliability and scalability for practical NLP applications. Despite these limitations, strong research efforts are focused on error correction and noise reduction, signaling optimism for viable quantum NLP solutions in the not-too-distant future.

Quantum computing's potential for NLP also extends into areas like language translation and interpretation, where it could dramatically increase accuracy and contextual understanding. Traditional machine translation methods can struggle with idiomatic expressions and disambiguation, often resulting in awkward or incorrect translations. The inherent capability of quantum systems to evaluate diverse possibilities concurrently might bring about improvements in these realms, resulting in translations that better capture the richness and nuances of human communication.

Another intriguing prospect is optimizing semantic analysis in vast corpuses of text. Quantum algorithms could revolutionize how we perform tasks like text classification, topic modeling, and sentiment analysis, by enabling a more nuanced understanding of the intricacies contained within human language. This would entail a leap forward in areas such as social media analysis, customer feedback interpretation, and automated content generation.

Furthermore, quantum computing can facilitate breakthroughs in unsupervised learning, a subfield of machine learning that is particularly pertinent to NLP. Unlike supervised learning, where models rely on labeled data, unsupervised learning deals with unlabeled data, seeking patterns and structures without explicit instruction. Quantum enhancement in unsupervised learning could allow for more sophisticated pattern recognition, deeper insights, and richer data analysis, which are crucial for dissecting the complexities of human language.

The fusion of quantum computing and NLP harbors transformative potential beyond conventional computing's reach. It's not just about faster processing times or handling larger datasets; it's about fundamentally reshaping the way we approach language tasks, paving the path toward advanced AI systems that comprehend context, sentiment, and meaning as seamlessly as humans. Such AI systems could have a profound impact on various industries, from customer service automation to content creation in media and entertainment.

In this evolving landscape, interdisciplinary collaboration is pivotal. Bringing together experts in quantum computing, linguistics, neural networks, and artificial intelligence is crucial for translating academic breakthroughs into real-world applications. The intersection of these fields holds the key to unlocking the mysteries of human language in ways previously unimaginable.

In conclusion, while the journey to fully integrating quantum computing with NLP is laden with technical hurdles, the possibilities are enticing. We stand on the brink of a new era where quantum NLP could potentially revolutionize how we interact with machines, making conversations more natural, intuitive, and meaningful. As research progresses and quantum technologies mature, we edge closer to a frontier where the full potential of language processing is realized, driven by the immense power of quantum computing.

Language Translation and Interpretation

In recent years, advances in natural language processing (NLP) have brought remarkable transformations to the fields of language translation and interpretation. As we continue to navigate a globally interconnected world, the demand for precision in linguistic conversion has never been more critical. The integration of AI-powered tools and quantum computing holds promise for overcoming many existing barriers, refining the capabilities of machines to understand and translate human languages with heightened accuracy. This section unravels the profound potential these technologies wield in reshaping communication across linguistic boundaries.

The road to sophisticated language translation technology has been long and winding, marked by incremental advancements in rule-based systems, statistical methods, and deep learning techniques. AI has empowered translation services by enhancing their capabilities through complex neural networks that can grasp context, idioms, and cultural nuances. However, errors still arise, particularly with languages featuring rich morphologies or ambiguous structures. Contextual understanding, subtleties in dialect, and idiomatic expressions pose challenges. Quantum computing is poised to revolutionize this arena by managing the inherent complexities and processing vast language datasets significantly faster than classical systems.

To understand the cutting-edge developments in language translation, one must first grasp the basics of NLP. Deep learning models, such as transformers, have been widely acknowledged for their role in improving accuracy and fluency in translations. These models, notably the self-attention mechanisms within them, allow for contextual awareness by processing sentences holistically rather than word-by-word. This advancement mimics how humans understand language—and that's where quantum computing steps in as a groundbreaking partner. Quantum algorithms promise to further expedite these processes, offering solutions that classical computers struggle with due to their limited processing capacity.

Quantum computers operate on qubits, which thanks to superposition and entanglement, process information simultaneously. This multitasking ability outshines binary systems and holds the capability to enhance NLP models' efficacy in translation operations.

Consider a scenario in which a quantum-enhanced NLP system analyzes a complex sentence; it could instantly evaluate myriad translation possibilities, choosing the most contextually and semantically correct option. This means faster, more nuanced translations, and thus improved real-time interpretation services.

Moreover, high-dimensional quantum spaces facilitate an exploration of what's called "word embeddings" in NLP—mathematical representations of words based on their meanings. By efficiently mapping semantic distances between words in different languages, quantum computing can help create translation paradigms that are richer in context and more attuned to the language's inherent qualities. This enables the translation of subtle expressions and complex idioms that traditional models might misinterpret or overlook.

Aside from the sheer speed and efficiency, quantum computing also brings potential breakthroughs in context preservation and ambiguity resolution. A significant hurdle in language translation is preserving the intended meaning, especially when the original content is subject to interpretation. Quantum NLP models are being developed to comprehend these nuances, disambiguate content, and choose translations that retain the deeper meaning of the source text. This aligns closely with the human approach to language comprehension, leading to outputs that are not only accurate but also resonant with the speaker's intent.

With these advancements, the field of interpretation—real-time translation during live dialogues or broadcasts—is set to evolve profoundly. Current AI-driven interpretation systems, while innovative, can still fall short by providing inaccurate translations or missing the speaker's emotional undertones. Quantum-enhanced AI systems could process vast streams of data in real time, ensuring that the linguistic and emotional content of a conversation is conveyed with greater fidelity. This opens new dimensions for cross-cultural communications, enabling seamless interactions across different languages.

The importance of privacy in such sensitive technological developments cannot be overstated. As language translation systems retain and process vast amounts of personal and sensitive data, ensuring security becomes paramount. Here again, quantum technologies are at the forefront, offering quantum encryption methods that bolster security and privacy. This assures users that their data remains protected while benefiting from advanced translation technologies.

In essence, language translation and interpretation are on the brink of a significant transformation fueled by the synergistic capabilities of AI and quantum computing. These technologies are interconnected, and their co-evolution will herald a new era of linguistic understanding and connectivity. The world stands at the precipice of a future where language barriers become relics of the past, and this transformation is set to redefine global collaboration and intercultural exchanges.

However, the journey is not without its obstacles. The challenges of integrating these advanced technologies into practical applications involve high resource demands, the current infancy of quantum hardware, and the need for extensive collaboration across industries and nations. But as these hurdles are overcome, language translation and interpretation will surely become more profound and transformative than ever imagined.

Looking ahead, the convergence of AI and quantum computing in NLP suggests endless possibilities. We can envision a landscape where language differences no longer impede communication, a reality where multilingualism dissolves geographical boundaries and fosters global harmony. This exploration of language translation and interpretation through the lens of emerging technologies heralds an age where every voice is heard, and understanding transcends words, enabling a truly interconnected world.

Future Prospects for NLP Technologies

As we stand on the frontier of technology, the prospects for Natural Language Processing (NLP) are rapidly expanding. Over recent years, NLP has moved from a niche application to a powerful tool used in various industries—from healthcare to finance and entertainment. With each advance in underlying technology, like quantum computing and artificial intelligence, the potential for NLP to transform how we interact with machines grows exponentially.

One of the most promising aspects of future NLP technologies lies in enhanced language understanding and generation. Currently, NLP systems can perform astonishing feats in language translation, summarization, sentiment analysis, and more. However, there is significant room for improvement, especially in understanding context and nuance at a human level. Future technologies could enable systems to grasp colloquialisms, dialects, and even the subtleties of humor or sarcasm more effectively than ever before. As AI models evolve through enriched data and quantum-enabled computational power, these capabilities are more within reach than ever.

The integration of quantum computing offers a tantalizing prospect for the optimization of NLP algorithms. Quantum computers promise to solve certain types of computational problems more efficiently than classical machines. In NLP, this could mean processing vast datasets at unprecedented speeds, allowing models to learn with more granularity and adaptability. Imagine a voice-activated assistant that understands your requests with such precision and speed that it feels almost human—this might soon be possible.

Language translation might experience a revolution, moving beyond mere bilingual exchanges to multilingual fluency across global languages, including less widely spoken languages. The capacity to efficiently process extensive linguistic data through quantum computing could even bridge cultural gaps, enabling smoother, more accurate, and more meaningful communication between people of varied linguistic backgrounds. With rapid advances, we might soon converse effortlessly across languages, almost as if language barriers never existed.

Furthermore, the expansion of NLP technologies can play a critical role in multilingual education. Future applications could provide personalized learning experiences based on an individual's language proficiency and learning pace. Such tools could level the educational playing field, offering individuals worldwide access to high-quality education materials, regardless of the language they speak. These prospects align with global aspirations of accessibility and inclusivity in education, potentially transforming how knowledge is disseminated worldwide.

The seamless integration of NLP with other AI technologies, such as machine learning and neural networks, heralds a new era of smart assistants. Imagine systems that not only understand what you say but anticipate your needs, adapting to your preferences with minimal input. These advances could redefine productivity tools, enabling more intuitive interfaces with technology for everyday users.

In customer service, NLP-powered systems could redefine the user experience by delivering more nuanced and empathetic interactions. As technology evolves, customer service bots might no longer provide generic responses but offer tailored solutions naturally, enhancing customer satisfaction significantly. The potential cost savings for businesses, combined with improved services, provide a clear incentive for investing in these future technologies.

Another exciting prospect is the application of NLP technologies in mental health and wellness. With future NLP systems capable of nuanced emotional understanding, they could assist in identifying emotional cues and offering preliminary counsel, thus supporting mental health professionals in delivering more effective care. This dual role of assisting humans while expanding our capacity to understand human emotions and language interactions signifies remarkable progress.

The entertainment industry, especially the realms of gaming and interactive storytelling, stands to benefit immensely from advanced NLP technologies. By augmenting real-time dialogue and interaction with players, the boundaries between virtual and reality could blur, creating experiences that feel genuinely immersive. Characters could respond to player's inputs with contextual understanding, making each interaction unique and engaging.

Data privacy will be a significant consideration as we push the boundaries of what NLP can achieve. Protecting user data while optimizing NLP models can become a delicate balancing act. Future developments must ensure robust security measures are in place to safeguard personal information, aligning technological advancements with ethical considerations. This will be paramount in gaining public trust and ensuring widespread adoption of advanced NLP technologies.

Concurrently, we may witness the rise of new business models and industries centered around advanced NLP capabilities. As these technologies mature, they can drive innovation across sectors, requiring professionals to build expertise in new fields. This interdisciplinary approach can foster collaboration between linguists, psychologists, data scientists, and technologists, enriching each domain with insights from others.

The future also holds potential regulatory challenges. As Conversational AI becomes more lifelike, distinguishing between human and machine-generated interactions will require careful consideration. Ensuring transparency, ethical AI usage, and accountability will be key to developing responsible NLP technologies that align with societal values. Robust policy frameworks and international cooperation could serve to guide these innovations in a morally responsible manner.

In essence, the future of NLP technologies is one of constant evolution, driven by ongoing breakthroughs in quantum computing, AI, and machine learning. The possibilities are practically limitless, with implications for numerous aspects of daily life and industry. By harnessing these advancements, we can look forward to a time where technology enhances our linguistic interactions, making communication faster, clearer, and more meaningful than today. The journey ahead holds not only technical challenges but immense promise, as we strive to bridge language divides and achieve a deeper connection across the global community.

Chapter 13: Ethical Considerations in Quantum AI

As quantum computing and artificial intelligence intersect to forge unprecedented technological advancements, ethical considerations become pivotal. The potential of Quantum AI to reshape industries and solve complex global challenges is immense, yet it equally demands a framework anchored in responsibility and fairness. Understanding and addressing biases within AI systems are vital, given the profound impact these technologies can wield on society's trust and equity. Moreover, as we leap forward, setting clear policies and regulations will be crucial in ensuring that Quantum AI's evolution aligns with societal values. This chapter navigates these ethical terrains, emphasizing the importance of foresight in developing systems that not only advance human capability but also uphold the principles of inclusivity and justice in an increasingly quantum-powered world.

Responsible AI Development

In the landscape of evolving technologies, the notion of responsibility cannot be overstated when it comes to the development of Quantum AI. At its core, responsible AI development seeks to ensure that advanced technologies are not only powerful but also aligned with human values, ethics, and societal well-being. As quantum computing and AI increasingly merge to form Quantum AI, the stakes are raised. This powerful intersection has the potential to revolutionize industries and solve complex global challenges, but it can also exacerbate inequities and create new ethical dilemmas if not pursued responsibly.

Quantum AI stands poised to redefine the benchmarks of computational power and efficiency. However, this very potential demands a cautious approach where ethical guidelines are as integral to the process as technical advancements. The process of responsible AI development involves several critical facets, including transparency, accountability, inclusivity, and fair governance. Organizations and developers must commit to these principles to prevent misuse and unintended consequences of Quantum AI applications.

One critical aspect is the transparency of algorithms and decision-making processes. In traditional AI systems, understanding algorithmic decisions poses a challenge, and with Quantum AI, this complexity multiplies. Developers should ensure that Quantum AI systems can explain their decisions in a manner that is comprehensible to humans. This requires building mechanisms that not only predict outcomes but also describe the variables influencing those predictions. Such transparency is vital for fostering trust among users and stakeholders.

Accountability is another cornerstone of responsible AI development. It's essential for developers and organizations to take responsibility for the systems they create. This includes maintaining oversight throughout an AI system's lifecycle—from design and implementation to deployment and monitoring. By fostering a culture of accountability, they can ensure that AI systems are used as intended, thus minimizing harm and enhancing trust in Quantum AI technologies.

Moreover, inclusivity in AI development helps to ensure that Quantum AI systems cater to the diverse fabric of society. As developers, making deliberate efforts to include a broad spectrum of voices during the design process can ensure that these systems do not perpetuate existing biases. An inclusive approach not only leads to more equitable technologies but also inspires innovation by integrating multiple perspectives and experiences.

Ethical governance frameworks are also vital for responsible AI development. These frameworks guide the ethical deployment and use of Quantum AI systems. They need to be dynamically evolved and iteratively tested to adapt to the fast-paced advancements in technology. Collaborating with ethicists, technologists, and policymakers can help frame these guidelines in a way that is both effective and versatile. Such collaboration will lay down a strong ethical foundation that aligns technological growth with societal values.

As AI systems become more autonomous and powerful with quantum capabilities, ethical considerations surrounding data privacy and security mount. Quantum technology's potential to crack encryption codes poses a conundrum that demands profound ethical scrutiny. Safeguarding data privacy in a way that respects individual rights while taking advantage of quantum capabilities is an essential component of responsible AI development. This involves crafting new cryptographic methods and protocols that are resilient against potential Quantum AI intrusions.

The pace of Quantum AI development also calls for proactive education and dissemination of knowledge regarding its ethical implications. By spreading awareness and educating both developers and the public about potential risks and benefits, a more responsible approach toward the technology's deployment can be cultivated. This education shouldn't be static; it must evolve alongside technological advancements, ensuring that responsible AI development remains robust and relevant.

Additionally, collaboration across borders and disciplines is a strategic asset in the mission for responsible AI development. Sharing knowledge and experiences globally can lead to more comprehensive understanding and management of ethical challenges. International cooperation can help create universal standards and practices that govern the ethical deployment of Quantum AI, balancing innovation with societal norms and security.

Quantum AI is not simply a tool for transformation; it holds the mirror to human intention. It's an opportunity for the tech community to reflect on how technology could be wielded as a force for good. Through responsible AI development, stakeholders can harness Quantum AI not only to advance industry but also to enhance human life, equity, and sustainability.

In this context, foresight becomes crucial. Technologists and ethicists must collaborate to anticipate potential future scenarios and their ethical implications before they materialize. This forward-thinking approach requires envisioning the long-term impacts of Quantum AI on society, including potential disruptions in labor markets, shifts in socio-economic frameworks, and changes in educational landscapes.

It's paramount for developers and policymakers to strike a balance between technological ambition and ethical restraint. While Quantum AI's promise is vast, its responsible integration into human life must be cautiously pursued. In meticulously aligning Quantum AI development with human-centric values, stakeholders can ensure that the profound capabilities of these technologies are harnessed responsibly and sustainably. This chapter of responsible AI development seeks not only to guide but to inspire a future where Quantum AI is a testament to thoughtful innovation, a future shaped by purpose as much as by possibility.

Addressing Bias and Discrimination

As we delve deeper into the potential of quantum AI, it becomes imperative to address the ethical concerns that may arise alongside its development and deployment. One of the most significant issues within this realm is bias and discrimination, phenomena not only ingrained in traditional AI systems but likely to persist, or even intensify, in quantum AI if not proactively managed. In AI systems, biases can emerge from the data they're trained on, often reflecting historical inequalities or prejudices. The danger lies in perpetuating these biases, or worse, amplifying them, in quantum AI systems that have unprecedented processing power and decision-making capabilities.

Understanding the roots of these biases is essential. Data bias arises when the input to an AI system does not accurately represent the broader context it's meant to function within. Whether due to the socio-cultural context, the lack of diversity in training datasets, or the conscious and unconscious biases of developers, these issues can deeply embed discrimination within the fabric of AI systems. In the context of quantum AI, algorithms could work on a scale previously unimaginable, with biases manifesting more rapidly and across a broader spectrum of applications.

Furthermore, the unique architecture of quantum computing introduces new complexities in how we must address these biases. Quantum algorithms operate in ways that can be opaque even to their developers, leveraging superposition and entanglement to process diverse possibilities simultaneously. This 'black-box' nature complicates efforts to identify and mitigate biases because it challenges traditional debugging and auditing techniques used in classical AI systems. Thus, interdisciplinary approaches combining expertise from quantum physics, computer science, and social sciences are crucial to unveiling and understanding biases in quantum AI systems.

In tackling these ethical hurdles, transparency is an indispensable principle. Ensuring that quantum AI models are interpretable and their decision-making processes are open to scrutiny can mitigate potential biases. Researchers and developers must endeavor to document and make accessible the processes through which quantum systems arrive at decisions. In practice, this could mean developing standardized methodologies for auditing quantum AI systems, potentially leveraging techniques such as explainable AI (XAI) that offer insights into the decision-making processes of algorithms.

Addressing discrimination requires an active, ongoing engagement with diverse perspectives. It's imperative to involve stakeholders from various backgrounds in the development process of quantum AI—whether they are from marginalized communities, industry experts, policymakers, or end users. By incorporating broad and inclusive viewpoints, the quantum AI field can aim to build systems that not only pride themselves on technological prowess but also on fairness and equity.

Another strategy involves the careful curation and management of data. The data used to train quantum AI systems should be representative and free from inherent biases to the greatest extent possible. Overcoming this challenge often means developing innovative data collection techniques and actively seeking data from diverse and traditionally underrepresented groups. It can also involve applying data correction methodologies that work beforehand to remove biases from datasets, thereby making the quantum AI systems less prone to discrimination.

Moreover, developers and organizations should foster a culture of continuous responsibility and commitment to ethical considerations. It's not enough to address biases at the initial stages of system development; ongoing assessments and updates are vital as the technology and its applications evolve. Regular ethical reviews, updates to training datasets, and adaptations to algorithms constitute proactive measures to identify and rectify bias before it becomes entrenched in decision-making processes.

Quantum AI policies and frameworks should explicitly define and enforce standards for anti-bias practices. These policies can set expectations for transparency, data integrity, and fairness, providing guidance for developers and companies in what they should uphold in both research and practical deployment. Furthermore, these frameworks should be adaptable to the rapid advancements in the field, allowing for flexibility in addressing novel ethical challenges as they arise.

Engaging with policymakers and regulatory bodies is crucial for establishing legal frameworks that support anti-discrimination efforts in quantum AI. As quantum computing capabilities expand, legal protections must adapt to ensure they encompass all aspects of AI and computing technologies. Collaborative efforts between technologists and legal experts can lead to coherent regulations that safeguard against bias while fostering innovation.

The scholarly community should also prioritize research dedicated to understanding how bias and discrimination manifest in quantum AI contexts. This includes studies focused on the interaction between quantum computing paradigms and the social dimensions of technology deployment. Academic partnerships and funding initiatives directed at these pressing issues can accelerate the development of tools and techniques to combat bias.

While challenges in addressing bias and discrimination in quantum AI are formidable, the opportunities for meaningful progress are equally vast. As quantum computing ventures further into strategic sectors, establishing ethical standards that emphasize equity and justice becomes non-negotiable. These efforts require concerted action, equal parts innovation, introspection, and collaboration, to ensure that the future of quantum AI transcends mere technical achievements and aligns with societal values that prioritize human dignity and fairness.

Quantum AI Policies and Regulations

As we stand on the brink of a quantum revolution, the urgency to establish comprehensive policies and regulations for Quantum AI becomes paramount. Harnessing the immense potential of Quantum AI to redefine industries and address global challenges must be paralleled by frameworks that ensure ethical deployment and equitable access. In a landscape where technology progresses at a bewildering pace, regulations are often playing catch-up, necessitating a proactive approach to guard against misuse and unintended consequences.

The first step towards constructing such a regulatory canvas involves understanding the unique attributes of Quantum AI. Unlike classical AI systems, this technology intertwines with quantum computing, bringing extraordinary computational prowess and a wholly distinct set of challenges. Superposition and entanglement, the cornerstones of quantum computing, introduce complexities that transcend classical regulatory approaches. Policymakers must delve into these intricacies, crafting regulations that reflect the nuanced nature of Quantum AI while encouraging innovation.

One crucial aspect that regulations must address is the potential for unprecedented data processing capabilities. Quantum AI could significantly enhance data analysis, leading to breakthroughs in fields such as healthcare, finance, and beyond. However, this capability raises concerns about data privacy and security, particularly when dealing with sensitive information. Regulatory frameworks need to address these concerns head-on, emphasizing transparency, accountability, and robust data protection measures to prevent misuse of data.

Moreover, as Quantum AI becomes a viable tool for both innovation and surveillance, the ethical implications of its deployment in various domains must be thoroughly evaluated. The advent of AI-driven surveillance techniques, empowered by quantum speed, could have profound societal implications. Governments and organizations should implement regulations that balance national security interests with individuals' rights to privacy, ensuring that surveillance is justified and proportionate.

International cooperation is another pivotal element in the global regulatory framework for Quantum AI. The seamless flow of data and technology across borders necessitates a coordinated effort to standardize policies and sharing of best practices. Collaborative international initiatives can help harmonize regulations, minimizing conflicts and fostering a cohesive approach to harnessing Quantum AI's potential. Organizations like the United Nations and the World Economic Forum could play a catalytic role in championing global standards for ethical AI use.

However, global cooperation shouldn't come at the expense of accommodating cultural and regional nuances. Regulations must be flexible enough to address local contexts while aligning with broader global objectives. This balance would allow countries to leverage Quantum AI in ways that are most beneficial to their unique socio-economic landscapes, ensuring that no nation is left behind in this technological leap.

Another fundamental element of Quantum AI policy-making is managing the dual-use nature of this technology. While it can drive positive changes in medicine, climate science, and more, it also zips along with the potential for less benign applications, such as advanced military weaponry or sophisticated cyber-attacks. Therefore, policies need mechanisms to anticipate and mitigate risks associated with dual-use technologies, constructing barriers to unethical applications while fostering legitimate innovation.

In setting policies and regulations, stakeholders must prioritize inclusivity and prevent inequality. Quantum AI holds transformative promises, but if benefits are not equitably distributed, it could widen existing socio-economic divides. Policymakers should focus not only on incentivizing the growth of quantum technology sectors but also on creating accessible educational pathways and ensuring equitable tech distribution. This focus will empower diverse population segments to contribute to and benefit from advancements in Quantum AI.

To effectively implement these regulations, there is a need for well-defined governance structures. Regulatory bodies at regional, national, and international levels should be established or empowered to govern the nuances of Quantum AI use. These bodies need clear mandates, responsive legislative tools, and accountability mechanisms to adapt to the fast-paced evolution of technology. A dynamic governance model, possibly utilizing quantum-enhanced AI for real-time oversight and decision-making, could offer a path forward in this rapidly advancing era.

The journey towards comprehensive Quantum AI regulations also requires a cultural shift in how we perceive innovation and risk. As stakeholders—scientists, policymakers, and the public alike—understand more about Quantum AI, a more mature dialogue can emerge that moves beyond fear-mongering to embrace informed risk-taking. Fostering this dialogue can encourage an ethical mindset in the developers and deployers of Quantum AI models, embedding ethical considerations into the very fabric of innovation processes.

Equally, academic institutions have a vital role to play. By integrating ethical discussions within quantum and AI curricula, they can cultivate a generation of technologists who not only understand the mathematical and scientific principles at play but also appreciate the ethical and societal dimensions of their work. Such interdisciplinary educational paradigms can prepare future talent to engage with regulatory discussions critically and creatively.

Creating a regulatory ecosystem that's both dynamic and resilient will require perpetual iteration. Policymakers need to remain agile, continuously evaluating the efficacy of regulations, assessing emerging trends, and engaging with diverse stakeholders. Public consultations, expert panels, and collaborative industry initiatives can offer invaluable insights into this iterative process, ensuring that regulations remain relevant and forward-looking.

Ultimately, the challenge of regulating Quantum AI is reminiscent of navigating uncharted territories. Yet, by championing ethics, fostering global cooperation, cultivating resilience, and embracing inclusion, societies can harness the power of Quantum AI responsibly. This approach doesn't just buffer against potential pitfalls but also unlocks vast opportunities to transform global challenges into achievements, ensuring that Quantum AI becomes a force for good in an interconnected world.

Chapter 14: Quantum AI in Education

As we continue to explore the transformative power of quantum AI across industries, its potential in revolutionizing education cannot be overstated. Quantum AI holds the promise of crafting personalized learning experiences tailored to each student's needs and pace, thereby enhancing educational outcomes in ways previously unattainable. This technological confluence can develop AI-driven educational tools that adaptively upgrade as a student progresses, offering real-time feedback and fostering curiosity and engagement. Additionally, integrating quantum computing into the curriculum equips learners with the knowledge of cutting-edge technologies and prepares them for future innovations. The fusion of quantum computing principles and AI in educational frameworks is set to not only democratize learning but also nurture the next generation of thinkers, ready to tackle global challenges with ingenuity and creativity.

Personalized Learning Experiences

In today's digital age, the notion of a one-size-fits-all educational model is rapidly giving way to more tailored, individualized approaches. The integration of Quantum AI in education holds transformative potential in creating personalized learning experiences that cater to the unique needs of each learner. By leveraging the immense processing power of quantum computing along with the adaptability of AI, we can begin to address the limitations of traditional education systems and move towards a future where learning is as unique as each student.

Quantum AI enables the rapid analysis of vast amounts of educational data, allowing for real-time adjustments to teaching methods and materials. This is made possible by the symbiosis between quantum computing's processing capabilities and AI's pattern recognition and decision-making prowess. For instance, consider a digital learning platform powered by Quantum AI that continuously assesses a student's learning style, pace, strengths, and weaknesses. Such a system could dynamically recommend resources and learning paths that are most effective for the student, thus optimizing the learning process.

At the heart of this personalization is the concept of adaptive learning. Unlike traditional systems that provide a static curriculum, adaptive learning systems continuously tailor educational content to match the student's evolving needs. Quantum AI can process these adjustments in real-time, accounting for complex variables such as a student's emotional state or specific topic comprehension. This leads to more efficient learning, with students achieving mastery more quickly than in a standard environment where each student follows the same path.

The personalization extends beyond merely academic performance. Quantum AI in education also has the potential to influence how students engage with their learning environments. By analyzing behavioral data, such systems can create a more engaging, motivating learning experience, utilizing gamified elements or immersive virtual reality scenarios that resonate with individual preferences and interests. As students receive instruction in ways that capture their attention and imagination, the often-daunting gap between student capabilities and curricular expectations begins to close.

Moreover, Quantum AI has a role in promoting equality in education. By analyzing vast datasets from diverse backgrounds, biases that may exist in traditional educational practices can be identified and corrected. The result? More equitable access to quality learning experiences, regardless of geographic, economic, or social differences. With the ability to adapt to multiple languages and cultural contexts simultaneously, Quantum AI-powered systems can support a global audience, promoting inclusivity and breaking down barriers that have traditionally segregated educational opportunities.

In practice, implementing personalized education through Quantum AI will require the development and integration of innovative technologies into current educational frameworks. Educators will need training and support to shift their roles from traditional instructors to facilitators of personalized learning experiences. While this transformation poses several challenges, the benefits of a more engaged, motivated, and successful student body are undeniable.

The transformation is already underway, with institutions experimenting with pilot programs that integrate AI-driven personalization into their teaching methods. As these technologies mature, they will offer granular insights into student learning patterns, recommend timely interventions, and propose customized challenges to keep students on the optimal learning trajectory. These interventions will ensure that no student is left behind, as each learning journey is closely monitored and adjusted.

It's important to recognize that the role of teachers will become more crucial than ever. Even the most advanced Quantum AI systems cannot replace the human aspects of teaching—empathy, encouragement, and inspiration. Instead, these tools will equip educators with detailed, actionable insights that allow them to focus on interpersonal aspects, guiding and supporting students through their educational journey with greater efficacy.

In essence, Quantum AI's contribution to personalized learning is not about creating isolated, tech-driven encounters but rather supporting enriching human connections facilitated by data-driven insights. As students take center stage in their learning paths, educators can function more effectively as guides and mentors, nurturing each student's potential and creativity.

Looking forward, the implications of personalized learning experiences driven by Quantum AI are vast. This progression promises not only to reshape education by making it more accessible and efficient but also to equip learners with the skills and critical thinking required in a dynamically changing world. Each student, irrespective of their original starting point, can achieve a level of mastery previously deemed unattainable, thus fulfilling a fundamental promise of education—to enlighten and empower individuals to pursue limitless possibilities.

Ultimately, as Quantum AI continues to evolve, its capacity to revolutionize the educational landscape becomes clearer. The journey towards fully personalized learning is not without its hurdles, but with continued innovation and thoughtful implementation, the dawn of a new era of education is within sight, promising to craft a future where learning is tailored to unlock the potential of every student.

AI-driven Educational Tools

The advent of AI-driven educational tools marks a transformative epoch in the sphere of learning and teaching. As we stand on the cusp of a quantum AI revolution, education emerges as one of the most promising fields, ready to be reshaped by the synergies between quantum computing and artificial intelligence. These tools carry the potential to redefine our conception of learning, steering it toward a more personalized, efficient, and accessible paradigm.

At the core of these AI-driven tools is the ability to cater to the individual needs of each learner, moving away from the traditional one-size-fits-all approach. With AI's capability to analyze vast datasets, educational tools can now adapt to the specific pace, style, and preferences of each student. This leads to a learning experience that feels tailor-made, helping students grasp complex concepts effectively and at their own pace. The rise of personalized learning paths promises not only improved educational outcomes but also heightened student engagement.

AI-driven tools are particularly powerful in interpreting and predicting students' learning behaviors. This predictive capability allows educators and institutions to identify potential challenges students might face even before they become obstacles. For instance, AI algorithms can recognize patterns suggestive of a student's struggle with particular topics, prompting timely interventions from educators. This proactive approach ensures that learning difficulties are addressed promptly, minimizing the risk of student frustration or dropout.

Moreover, the integration of quantum computing into these AI systems is set to heighten their performance and efficacy. Quantum algorithms bring unparalleled computational power, allowing educational tools to process and analyze data at speeds and scales that are unattainable with classical computers. This means that educational AI can make more informed, accurate predictions and adaptations, enhancing the precision of personalized learning experiences.

Beyond personalization, AI-driven educational tools are redefining the role of educators as well. By automating administrative tasks such as grading and tracking student progress, AI frees up educators to focus more on instruction and student interaction. This shift allows teachers to become more like facilitators and mentors, guiding students through their personalized learning journeys and nurturing critical thinking and problem-solving skills.

Furthermore, AI-driven educational tools are democratizing education, making high-quality learning resources accessible to a wider audience. These tools have the potential to transcend geographical and socioeconomic barriers, bringing education to remote and underserved communities. Online platforms powered by AI can offer interactive and comprehensive courses, reaching students who might otherwise lack access to quality education and opening up opportunities for lifelong learning.

Another significant impact of AI-driven tools is their ability to provide real-time feedback. Traditional educational systems often rely on periodic assessments, which might delay feedback. With AI, students can receive immediate insights into their performance, allowing them to adjust their learning strategies promptly. This immediate feedback loop is crucial in fostering an environment of continuous improvement and motivation.

Integration of AI in education also fosters collaboration and interaction among students from diverse backgrounds. AI-powered platforms can facilitate group projects and discussions, promoting a global learning community. These interactions enhance cultural awareness and empathy, preparing students to thrive in an increasingly interconnected world.

Of course, while the prospects are promising, implementing such advanced technologies in educational settings comes with challenges. There are concerns about data privacy, the potential for bias in AI algorithms, and the digital divide between students with varying levels of access to technology. Addressing these issues requires careful consideration and robust frameworks to ensure that AI in education remains a force for good.

As we look to the future, the ongoing integration of quantum AI in education is expected to bring forth even more innovative tools and solutions. Envision a classroom where quantum-enhanced AI systems can simulate complex scientific phenomena or historical events, offering students immersive and experiential learning. These tools could revolutionize subjects traditionally perceived as challenging, like physics and chemistry, making them more accessible and engaging for all students.

In conclusion, AI-driven educational tools hold immense potential to reshape the educational landscape. By leveraging the capabilities of both AI and quantum computing, these tools promise to make education more accessible, personalized, and efficient. As we continue to develop and refine these technologies, it is essential to do so with an ethical framework that safeguards equity and inclusion, ensuring that the benefits of AI-driven education are enjoyed universally. The journey toward this vision is an exciting one, promising not only advancements in how we learn but also in how we live and work in a rapidly transforming world.

Quantum Computing as a Curriculum

In the rapidly evolving landscape of education, the introduction of quantum computing as a curriculum holds transformative potential. The pursuit of understanding quantum mechanics and its application to artificial intelligence is not simply an academic exercise; it's the foundation of tomorrow's technological paradigms. As we stand on the brink of a quantum revolution, integrating these subjects into educational structures can prepare future generations to harness and innovate with quantum technologies.

Quantum computing differs starkly from classical computing. It operates on the principles of quantum mechanics, using qubits instead of bits. Bringing this cutting-edge concept into classrooms requires educators to not only familiarize students with the abstract nature of quantum mechanics but also to instill a mindset oriented towards exploration and problem-solving in multidimensional spaces. This shift could redefine educational methodologies, moving away from rote memorization to a focus on critical thinking and innovation.

Incorporating quantum computing into the curriculum doesn't mean starting from scratch. The basic tenets of STEM education—Science, Technology, Engineering, and Math—can serve as a scaffold upon which new ideas are built. Adding layers of quantum theory and its practical applications can fundamentally enrich this existing framework. Schools and universities that adapt quickly to these changing educational demands can position themselves as leaders in modern, forward-thinking education.

As students delve into qubits, superposition, and entanglement, they also benefit from the intersection of quantum computing with artificial intelligence. This nexus offers a fertile ground for nurturing groundbreaking ideas. Introducing quantum algorithms and their specific advantages can provide students with the tools needed to harness quantum AI in addressing real-world challenges, from complex data analysis to advanced cybersecurity measures.

Furthermore, the curriculum must reflect current research and advances in the field. In this regard, collaboration with leading institutes and participation in cutting-edge research can provide students with invaluable exposure to the dynamic nature of quantum technologies. Creating a curriculum that's adaptable to new discoveries ensures that students are not learning outdated information, but rather engaging with a subject that's very much alive and continually evolving.

Educators face the challenge of making complex and abstract concepts relatable. This can be achieved through project-based learning and interactive simulations that bring quantum phenomena into tangible, understandable experiences. By encouraging students to build and test models, they gain insights into the practical implications and limitations of quantum technologies. This hands-on approach can stimulate curiosity and foster a deeper understanding that goes beyond theoretical knowledge.

Developing quantum computing courses also involves addressing the misconceptions and barriers to entry often associated with the subject. Quantum topics might initially seem intimidating, shrouded in complexity and counterintuitive concepts. By demystifying these principles through accessible teaching methods, educators can make quantum studies more approachable. Bridging the gap between complex quantum ideas and beginners' understanding should be a priority.

The inclusion of quantum computing in education should also highlight ethical considerations. As students learn about the immense potential and transformative power of quantum computing, they should also engage in discussions about ethics, societal impacts, and the responsibility that comes with such powerful technologies. Examining these dimensions instills a sense of ethical responsibility and prompts students to critically consider the broader implications of their work.

Quantum computing courses need to incorporate interdisciplinary collaboration as well. By intertwining fields such as computer science, physics, and even philosophy, the curriculum can offer a holistic education that prepares students to think across disciplines. Encouraging partnerships between different academic faculties can ensure a well-rounded educational experience, fostering innovation and a more comprehensive understanding of the quantum ecosystem.

Moreover, partnerships with tech companies and research facilities can provide practical experiences and mentorship opportunities. These connections offer students a glimpse of how quantum computing is used in industry and research, allowing them to see the practical applications of their studies. Internships, collaborative projects, and guest lectures can significantly enrich the curriculum and enhance learning outcomes.

One can't overlook the necessity of preparing teachers as well. Professional development for educators is crucial for the successful implementation of a quantum computing curriculum. Comprehensive training programs and resources must be established to equip teachers with the necessary skills and knowledge required to instruct in this complex field. Investing in teacher training ultimately benefits students, as it improves the quality and delivery of quantum education.

The advent of quantum computing as a curriculum marks an exciting new chapter in education. As the demand for quantum literacy grows, academic institutions will play a critical role in shaping knowledgeable and skilled individuals capable of driving this frontier forward. By laying down a robust educational framework, we prepare the thinkers and innovators who will push the boundaries of technology and transform how we understand, interact with, and reshape our world.

Chapter 15: The Future of Robotics with Quantum AI

As we delve into the transformative synergy between robotics and quantum AI, the horizon of technological innovation appears boundless. Imagine robots, not just executing pre-programmed tasks, but learning and adapting in environments of unprecedented complexity, powered by the astounding processing capabilities of quantum computing. This integration allows for real-time decision-making with precision unimaginable in classical frameworks, revolutionizing industries from manufacturing to healthcare with unparalleled efficiency. Robotics, empowered by quantum algorithms, can suddenly grasp patterns in data that were once lost in noise, enhancing autonomous learning and interaction. The melding of quantum AI with robotic systems opens doors to applications across diverse domains, fostering an era where machines can more intuitively work alongside humans, leading to enhanced productivity and creative partnerships. Envision a future where robots, equipped with quantum insight, navigate chaos with clarity, adapting fluidly to emergent challenges, and its potential to reshape industries is as exciting as it is transformative.

Integrating Quantum Processing in Robotics

The integration of quantum processing into robotics marks a paradigm shift that redefines the potentialities of automated systems.

Combining the principles of quantum computing with the mechanical prowess of robotics offers a leap forward in terms of efficiency, precision, and capabilities that were previously unimaginable. At its core, quantum processing brings an enhanced capacity for problem-solving, allowing robots to operate in complex environments with greater adaptability. This fusion is about more than just upgrading computation speed; it's about empowering robots with a new form of intelligence.

Traditionally, robotics has heavily relied on classical computing models, which, despite their advancements, are fundamentally bounded by linear and binary operations. Quantum processing, however, introduces qubits—quantum bits that exist in multiple states simultaneously thanks to the principles of superposition and entanglement. This capability drastically enhances the computational power available for solving problems that involve vast permutations and complex variables. For instance, navigating a robot through an unpredictable, dynamic environment becomes not only feasible but also efficient with the application of quantum algorithms.

Consider industries where robotics plays a crucial role—such as in manufacturing, logistics, healthcare, and beyond. Integrating quantum processing allows robots in these sectors to carry out highly complex tasks, such as analyz *real-time environmental changes* or optimizing logistical routes with unprecedented accuracy and speed. Such advancements open up new possibilities for automation, where tasks that once required substantial human intervention can now be delegated to machines. Quantum processing fosters a form of *evolution* in robotics—moving from pre-programmed, deterministic actions to cognitive, decision-based functionalities.

Another fascinating aspect worth exploring is robot learning. Conventionally, machine learning in robotics has achieved notable successes, yet faces limitations in processing and adapting to extensive datasets in real-time. Quantum processing, armed with quantum machine learning algorithms, equips robots with the ability to rapidly learn from and adapt to vast data inputs—enhancing their decision-making processes. Imagine a robot capable of learning complex scenarios such as emergency disaster response or intricate surgical procedures, gaining knowledge from a multitude of prior instances with unmatched precision.

Quantum-enhanced robotics isn't just an application; it's an opportunity for leaps in innovation across countless sectors. On one hand, the capabilities this technology introduces pose a challenge to existing regulatory frameworks, technical standards, and best practices. On the other hand, as research propels forward, it is critical to consider the ramifications of these advancing technologies on the workforce, ethical dimensions, and societal paradigms. How do we ensure these robots, empowered with quantum processing, align with ethical mandates and societal expectations? This dialogue needs to be ongoing and integrated into every step of innovation.

The vision of quantum processing in robotics also extends to collaborative robotic systems, or cobots, which are designed to work alongside humans. In scenarios requiring high levels of synchronous cooperation between humans and robots—such as assembly lines or medical environments—quantum processing can ensure that cobots perform their tasks while seamlessly adapting to human actions. This partnership between quantum cobots and human workers could redefine productivity, efficiency, and safety standards across sectors traditionally plagued by automation challenges.

Moreover, the integration of quantum processes could redefine the very architecture of how robots are designed and built. Lightweight algorithms, enhanced sensor processing capabilities, and real-time adaptability will necessitate a rethinking of both hardware and software paradigms in robotics. The potential to eliminate bottlenecks caused by transistors' limitations invites a new age of miniaturized, efficient robotic systems that are not only faster but also more energy-efficient and sustainable.

While the foundations of merging quantum processing with robotics are promising, significant challenges remain. Developing a cohesive infrastructure that supports efficient quantum computation in real-time scenarios is paramount. Additionally, there is the need to keep improving error correction mechanisms and to address current scalability limits within quantum systems. These technological hurdles don't deter the future of robotics with quantum AI; rather, they invite innovation, encouraging collaboration across disciplines—from quantum physicists to roboticists and software engineers.

The integration of quantum processing into robotics doesn't just redefine technological capabilities; it reshapes our imaginations about what robots can achieve. By leveraging the distinctive strengths of quantum computing, we edge closer to realizing a world where robots act not only as tools or assistants but as intelligent collaborators equipped with a profound understanding of complex environments. As we venture further into this quantum era, the symbiosis of quantum processing and robotics holds the promise of a future where the boundaries of automation and intelligence continually expand, setting the stage for breakthroughs previously the realm of science fiction.

Enhancing Robot Learning Capabilities

The convergence of quantum computing and artificial intelligence holds the promise of significantly enhancing robot learning capabilities. This intersection is particularly exciting because it brings with it the possibility of overcoming existing challenges in robotic learning, making robots not only more intelligent but also more adaptable and versatile across various tasks.

In today's world, robots are primarily driven by classical computing principles to learn and execute various tasks. However, as the complexity of tasks increases, these systems encounter limitations, especially when dealing with high-dimensional data and decision-making under uncertainty. This is where Quantum AI steps in, offering a paradigm shift through computational advances that were once deemed impossible by classical systems.

The foundation of any intelligent system, whether naturally occurring or artificially created, is its capacity to learn and adapt from its environment. For machines, this means acquiring the ability to process vast amounts of information and develop algorithms that lead to suitable and timely responses. Quantum computing introduces a level of parallelism in processing potential solutions that classical computing can't match. This parallelism allows for the consideration of multiple solutions simultaneously, which drastically reduces the time needed for computations.

Quantum-enhanced learning models, when integrated into robotics, could lead to unprecedented advancements in areas like autonomous navigation and manipulation, where rapid learning from dynamic environments is critical. For instance, quantum algorithms can optimize the training of neural networks by helping escape local minima problems more efficiently, leading to more refined and efficient learning processes.

Moreover, robots equipped with quantum-enhanced AI systems could surpass current limitations in processing speed and problem-solving capabilities. This would allow them to perform complex tasks such as understanding human emotions better, enabling a more nuanced interaction between humans and machines. Imagine a robot counselor capable of interpreting subtle emotional cues to engage in meaningful support, learning from each interaction to improve its responses.

The adaptability of robots is another aspect set to transform with quantum computing. Current AI models support learning that can adapt to specific tasks after extensive training. However, they struggle with generalizing this learning to different environments or tasks without significant retraining. The probabilistic nature of quantum algorithms can facilitate this generalization ability, giving robots a level of dexterity in their learning processes that mirrors more closely the human capacity for adaptation.

More technically, quantum kernels and variational quantum circuits can be applied to robotic learning models to tackle non-linearity and intricate decision landscapes more robustly. These methodologies can give machines an edge by allowing them to navigate complex datasets, find hidden patterns, and develop strategies in unprecedented ways. Such capabilities not only enhance robot intelligence but also pave the way for developing robots capable of solving tasks autonomously and efficiently in unknown or unpredictable settings.

Let's consider the impact on industrial robotics. With quantum computing, robots could greatly enhance quality control in complex manufacturing processes, employing real-time data analysis to detect anomalies or suggest optimization strategies on the fly. This not only increases efficiency but also extends the robots' operational capabilities beyond predefined constraints.

Implementing quantum algorithms in robotic learning isn't devoid of challenges, though. The integration of quantum hardware with existing robotics platforms presents technical hurdles, from maintaining qubit coherence to managing error rates. Nonetheless, these challenges are fuel for continued research and development, driving innovation. The potential benefits, such as improved learning speeds and efficiencies, provide compelling incentives to overcome these obstacles.

As we push forward, the research community is hard at work developing hybrid algorithms that take advantage of both quantum and classical resources. In such systems, classical algorithms could guide overall strategy and infrastructure, while quantum algorithms handle the most computationally intense subproblems. This synergy promises a new epoch of robot design, where machines learn dynamically, adapting not just to their environment but to the increasing complexity of the tasks at hand.

The journey towards advancing robot learning with quantum AI is not merely a technological quest; it is an exploration of what it means for machines to learn, think, and decide. It opens up a dialogue on ethics, responsibilities, and the philosophical implications of creating entities capable of mimicking aspects of human intelligence. Such machines might someday serve as partners in solving global issues, from environmental management to healthcare, extending human capabilities in ways previously thought unimaginable.

In conclusion, enhancing robot learning capabilities with Quantum AI is a venture into uncharted territories of machine intelligence. It provides a vision of the future where robots are not just tools but evolved entities alongside humanity. As advancements continue, the blending of quantum computing with artificial intelligence in robotics promises to usher in a new era of innovation, pushing the boundaries of what is achievable and setting a foundation for a future where intelligent, adaptable robots coexist harmoniously with humans. This is a future that eagerly awaits the exciting potential of quantum-powered robotics.

Applications in Diverse Industries

The potential of implementing Quantum AI in robotics extends far beyond what most could envision. As these cutting-edge technologies converge, they open new avenues in industries that are continually seeking innovation. Let's explore how Quantum AI not only transforms these sectors but also blurs the lines between what is technically possible and conceptually groundbreaking.

In the automotive sector, Quantum AI stands poised to redefine how we conceive of transportation. Autonomous vehicles can be significantly enhanced by optimizing decision-making processes in real-time. Quantum processors, with their unmatched capabilities in handling complex computation, allow autonomous systems to simultaneously analyze vast data arrays from multiple sensors. This capability translates to more efficient navigation, superior obstacle recognition, and a higher degree of safety. As Quantum AI continues to evolve, the prospect of fully autonomous fleets operating with minimal human intervention becomes increasingly plausible.

The energy sector is another field ripe for transformation. The inherent complexity of managing power grids, especially with the integration of renewable sources, presents numerous challenges. Quantum AI offers unparalleled precision in modeling and predicting energy consumption patterns. By optimizing these systems, energy providers can minimize waste and improve sustainability outcomes. Moreover, Quantum AI can facilitate the management of microgrids, enhancing their stability and efficiency.

Healthcare innovation, too, stands on the brink of a revolution. Robotics integrated with Quantum AI has the potential to transform surgical procedures. Imagine a robotic assistant endowed with quantum-enhanced capabilities, performing surgeries with precision levels previously unattainable by human hands alone. This could lead to reduced recovery times and increased success rates in complex surgeries. Additionally, Quantum AI can enhance telemedicine by ensuring data is processed rapidly, providing medical practitioners with real-time insights and decision-support systems without the delay associated with classical processing.

Manufacturing, a backbone of global economies, also benefits significantly from Quantum AI-enhanced robotics. Precision in manufacturing is paramount; minor errors can lead to significant financial repercussions. Quantum AI can boost quality control mechanisms by analyzing production variables with great accuracy. Predictive maintenance systems could rely on this technology to foresee equipment failures much earlier than traditional AI systems, thereby reducing downtime and maintaining productivity levels.

In the realm of logistics and supply chain management, Quantum AI has the potential to revolutionize how goods are distributed worldwide. By optimizing route planning and load balancing in real-time, Quantum AI-driven robotics can enhance delivery efficiency and reduce fuel consumption. This can lead to significant cost savings and environmental benefits. Furthermore, in warehouse settings, Quantum AI can empower robotic systems to make autonomous decisions about inventory management, ensuring that stock levels are maintained and shortages are averted.

Construction, often considered a labor-intensive industry, presents unique challenges that Quantum AI-enhanced robotics can address. Automated construction systems powered by quantum processors can operate with enhanced accuracy, ensuring that structures are erected with minimal discrepancies. This precision contributes to reducing material wastage and labor costs. The use of Quantum AI also enables predictive modeling of construction projects, helping to anticipate and mitigate risks associated with dynamic construction environments.

In agriculture, Quantum AI will empower robotic systems to perform tasks such as planting, harvesting, and monitoring crop health with extraordinary efficiency. These systems can integrate complex datasets from soil sensors, weather forecasts, and satellite imagery to optimize agricultural practices. As climate change continues to impact global food security, adopting Quantum AI in agriculture offers a promising solution to enhance crop yields and ensure sustainable farming practices.

The defense industry, always incorporating cutting-edge technology, stands to gain significantly from Quantum AI integration. Robotic systems used for reconnaissance, surveillance, and even tactical maneuvers can leverage the computational prowess of quantum systems. This can lead to more effective decision-making in high-stakes environments where outcomes are time-sensitive. Moreover, Quantum AI can enhance cybersecurity within defense systems, protecting critical infrastructure from increasingly sophisticated cyber threats.

Retail and e-commerce can also harness Quantum AI for enhanced customer experiences. Quantum processors can process consumer behavior data at speeds unimaginable with classical computing, driving personalized marketing strategies and optimizing supply chains. Robotic systems in warehouses can dynamically adjust to inventory demands, aligning stock levels with predictive purchasing trends. The seamless integration of Quantum AI in retail posits a future where consumer needs are anticipated with remarkable accuracy.

Finally, the entertainment industry's potential integration of Quantum AI offers unique opportunities for innovation. Robotic systems, enabled by quantum processing, can create interactive entertainment experiences, leveraging real-time decision-making to customize content dynamically. Whether in gaming or virtual reality environments, Quantum AI-enhanced robots can offer users enriched experiences, characterized by unparalleled immersion and interactivity.

Quantum AI in robotics across diverse industries not only offers improved efficiency and innovation but also heralds a paradigm shift in exploring and expanding the possibilities of technological integration. The evolution from theory to tangible applications is becoming a reality, catalyzing a new era in industrial operations where limits are constantly tested and often surpassed.

Chapter 16: Quantum AI for Art and Creativity

In the evolving landscape of technology, Quantum AI stands poised to revolutionize the realms of art and creativity, unveiling new possibilities at the intersection of code and inspiration. By harnessing the unique capabilities of quantum computing, artists and creators explore uncharted territories where the unification of complex quantum states gives birth to unprecedented forms of digital artistry. Musicians delve into quantum compositions, crafting harmonies and patterns that traditional methods simply can't replicate, while visual artists exploit quantum algorithms to breathe life into vibrant canvases of dynamic, data-driven art. As AI and quantum computing converge, they're not just tools for creativity; they become collaborators, offering artists an expanded playground to experiment with novel techniques and modes of expression. Through these forward-thinking collaborations, we witness a renaissance where the boundaries of imagination and technology are delightfully blurred, promising an era of innovation that both honors and transcends the essence of human creativity.

New Frontiers in Digital Art

Welcome to a fascinating new era where art and technology converge like never before. The fusion of quantum computing and AI has ushered in a revolutionary phase in digital art, fundamentally altering the creative landscape. This intersection isn't just about utilizing cutting-edge technology for creation; it's about redefining the essence of creativity itself. In this realm, every brushstroke, pixel, and sound wave can be transformed and reimagined, expanding the horizons of what is possible.

In traditional art forms, the artist's imagination was confined by the tools and mediums available. Today, quantum AI breaks down these barriers, offering artists a limitless array of possibilities. Quantum computing's unique abilities—such as superposition and entanglement—enable the processing of complex algorithms, allowing the creation of intricate patterns and forms previously unimaginable. Paired with AI's capability to learn and generate in ways akin to human thought, this technology duo actively participates in the creative process, sometimes surpassing human intuition. You could say it's like having a collaborator with an unparalleled understanding of the abstract.

At the heart of this transformation is the role of quantum algorithms, which redefine how digital art is generated and experienced. They allow for the manipulation of vast datasets, giving artists access to a broader palette of colors, forms, and textures than ever before. Imagine an AI capable of understanding not just color theory, but the emotional tone that different colors can evoke when used in certain contexts or patterns. This capability extends to mimicking artistic styles or even blending multiple styles to form new, unique modes of expression.

Unlike conventional AI, which often relies on predetermined datasets and styles, quantum-enhanced AI systems can think "outside the box", generating unexpected and novel outcomes. This unpredictability introduces an element of surprise and spontaneity to the creative process, which many artists find exhilarating. It challenges them to consider new concepts and visuals that might not have otherwise existed. In essence, quantum AI doesn't simply assist artists; it provokes and inspires them to reach beyond their usual limits.

Furthermore, quantum AI facilitates new forms of collaboration. Artists no longer have to work in isolation. Instead, they can engage in dynamic interactions with algorithms that contribute to the artistic process. This results in unique co-creations, products of a symbiotic relationship between human intuition and machine calculation.

The experience of art consumption is also transformed through quantum AI. Virtual reality (VR) and augmented reality (AR), when intertwined with quantum AI, offer immersive experiences that react and adapt to viewers' emotions and actions in real-time. Imagine viewing a painting that morphs and evolves based on the emotions it detects from you, creating a deeply personal interaction. Such experiences are becoming plausible through the convergence of these advanced technologies.

Importantly, the art market is also undergoing a transformation. We see new business models emerging, driven by the production of unique, digital artworks that are impossible to replicate. Blockchain technology, when combined with quantum computing, could radically change how art is valued, bought, and sold, making the process more transparent and secure.

Moreover, there's a democratization of art creation and appreciation. Quantum AI can lower the barrier to entry for budding artists, providing intuitive tools for those without formal training to express themselves creatively. This potential for inclusivity opens the doors for more diverse voices in the art world, enriching the global tapestry of culture and expression.

Yet, we must address the implications of authorship and originality in this new landscape. How do we define the role of an AI that contributes creatively? Who owns the art produced—human or machine? While these questions challenge our existing frameworks, they also push us to evolve and adapt, much like art itself.

We are on the cusp of a thrilling frontier, where art is not just about static images on a canvas but about dynamic experiences crafted through the synergy of quantum computation and artificial intelligence. The challenge and the opportunity lie in navigating this new space responsibly and creatively, ensuring that while technology advances, the human element remains central to artistic expression. As we embrace these new frontiers, we redefine what it means to create, experiencing the joy and wonder of discovering new art forms that reflect the boundless spectrum of the human spirit.

Music and Quantum Compositions

The nuanced relationship between music and mathematics has long been a subject of admiration and exploration. As we stand on the cusp of a new technological era, the fusion of quantum computing and artificial intelligence presents an unprecedented opportunity to delve deeper into this symbiosis. In the realm of music, the intricate patterns that define compositions can now be unraveled and reimagined with the help of quantum algorithms, potentially transforming not just how music is composed, but also how it's experienced.

Quantum computing introduces a paradigm shift in music composition through its inherent ability to process information in superpositioned states, known as qubits. This capability allows for the exploration of myriad musical possibilities simultaneously. While traditional algorithms might be confined to a linear processing mode, quantum algorithms can consider countless compositional permutations at once, generating musical outcomes that could surpass human imagination.

Consider the potential for creating music where notes do not follow in traditional harmony but instead take advantage of quantum entanglement, arranging themselves in unpredictable yet aesthetically appealing sequences. With quantum computing, compositions aren't merely concoctions of pre-existing rules but can explore entirely new musical architectures. This enables composers to embrace randomness and uncertainty as core elements of their creative repertoire, generating pieces that could evoke unfamiliar emotions or rhythms.

Beyond the technical capabilities, the integration of AI within the realm of quantum computing provides another layer of sophistication. AI's potential to learn and predict allows it to adapt music compositions dynamically, offering real-time customization to suit listener preferences or emotional states. Imagine a composition that evolves as it's being heard, intertwining itself with the listener's current mood, environment, or even physiological responses. This personalized musical interplay could introduce novel experiences, deepening the connective tissue between music and human emotion.

The notion of AI-driven collaboration isn't merely speculative. Artists and composers are already leveraging AI to push the boundaries of creativity. With the added dimension that quantum computing brings, collaboration could transcend human and machine boundaries. AI could work in tandem with human creators, proposing intricate compositions that defy traditional harmony or rhythm rules, sparking inspiration for new genres and styles.

An intriguing prospect is the possibility of designing instruments that respond to quantum-based compositions. Such instruments, echo chambers for quantum computational processes, could realize sound in ways unachievable by traditional means. These futuristic instruments might translate quantum states directly into audible music, bypassing standard digital synthesizers and offering a wholly new auditory palette.

Another dimension where quantum compositions might find fertile ground is in healing and therapy. Music therapists increasingly utilize melodies to evoke relaxation or emotional releases. Quantum compositions could be tailored to resonate with the neurological patterns of the brain, potentially offering innovative therapeutic solutions. Imagine quantum-composed soundscapes that can adjust in real time to the listener's physiological responses, offering a harmonizing effect on mental and emotional well-being.

Furthermore, the influence of this revolutionary technology could extend into the realm of musical analysis and archiving. AI integrated with quantum computing could examine vast archives of musical compositions, identifying subtle patterns and elements that elude human analyzers. This level of analysis might not only catalog historical trends in music but reveal new interpretive insights or inspire contemporary creators.

The intersection of quantum computing and music composition offers immense potential, jumping beyond the confines of imagination toward a future brimming with unexplored auditory possibilities. Yet, this comes not without challenges. Questions about the essence of creativity, the value of human touch in art, and the ethical considerations surrounding AI-generated content become pertinent. As we negotiate this brave new world, it is crucial to foster dialogues that honor the interplay between human artistry and machine capability.

Ultimately, the fusion of music and quantum compositions invites us to reconsider the very fabric of creativity. It challenges long-standing notions of authorship, collaboration, and artistic expression, offering a tantalizing glimpse of a future where the music we hear isn't just a melody but a multisensory experience artfully woven through the fabric of advanced computational realms. This future beckons us to explore, innovate, and redefine what it means to create and enjoy music in an age dominated by quantum mechanics and artificial intelligence.

AI-driven Creative Collaborations

In the world of art and creativity, collaboration has always been a catalyst for innovation. Artists, designers, and creators gain new perspectives and ideas when they work together, merging talents to form something greater than the sum of its parts. With quantum computing and artificial intelligence (AI) entering the scene, a new era of creative collaborations is emerging—one that transcends traditional boundaries and explores previously uncharted territories.

The infusion of AI into the creative process doesn't simply mean emulating or enhancing human capabilities but also involves creating entirely new artistic expressions. AI-driven creative collaborations often begin with a complex interplay between human intuition and AI's computational prowess. This partnership allows artists to explore creative possibilities that were once unimaginable, generating forms and structures that challenge our definitions of art. AI can process massive datasets—far larger and more complex than a human could handle—and identify patterns that can lead to unexpected yet exquisite outcomes.

Consider the realm of visual arts. Here, AI algorithms can analyze countless works from different epochs and cultures, discerning styles, tones, and subtle nuances. When artists collaborate with AI, they gain access to these insights, thereby expanding their creative vocabulary. The creative process becomes less constrained by individual knowledge and time limitations, allowing for real-time exploration of infinite stylistic variations. Imagine using AI to mix impressionism with digital pixel art, or blending Renaissance detail with abstract concepts, crafting entirely new artistic genres.

Music, too, is experiencing a revolutionary shift. AI-driven algorithms can analyze and generate music at scales and dimensions previously unattainable. When composers engage with AI, they find themselves working with an entity that can provide both a treasure trove of historical musical knowledge and the potential for contemporary experimentation. Instead of being seen as a tool, AI becomes a co-composer, suggesting harmonies, structures, and rhythms that can inspire human musicians. This kind of creative symbiosis encourages the development of rich, novel soundscapes and reorganizes the boundaries of musical composition.

Integrating quantum computing into this landscape augments AI's capabilities, particularly with respect to processing speed and data analysis. Quantum-enhanced AI can sift through enormous datasets, unearthing deep connections and stylistic nodes that traditional computers might overlook. This newfound computational power facilitates the generation of sophisticated models and simulations that can help artists experiment with creative possibilities in real-time.

Imagine designing virtual environments using quantum-accelerated AI, one where designers create dynamic worlds with intricate ecosystems that evolve unpredictably yet artistically. These environments could adapt based on user interaction, creating a living canvas that's both a medium for art and a platform for immersive experiences. Artists can manipulate factors like light, texture, and form on an unprecedented scale, sculpting organic, interactive spaces that modify and grow as they are explored.

This revolution in creative collaborations extends beyond individual artists and into industries such as film, fashion, and design. Filmmakers can utilize AI-driven tools to plan cinematic sequences, automating laborious tasks like editing and rendering. In fashion, designers employ AI to generate patterns and experiment with materials, eco-friendliness, and scalability. The creative process becomes an iterative loop of inspiration, experimentation, and reflection—one that's dynamically reshaped by the insights AI provides.

The educational sector stands to benefit profoundly from these advances. By incorporating AI into art education, students can experience an inclusive and expansive learning environment. AI can offer personalized feedback, suggest creative pathways, and provide a tailored curriculum that adapts to individual needs and aspirations. This democratization of creativity democratizes access, making it feasible for learners from diverse backgrounds to explore and nurture their talents without creative bounds.

As we step into this brave new world of AI-driven creative collaborations, we also encounter important ethical questions. Who owns the art produced through AI collaboration? How do we attribute creativity and credit in a world where human and machine contributions are interwoven? These questions aren't just philosophical; they have tangible implications for licensing, copyrights, and the global market of creative works. Discussions around ethics and ownership will need to evolve alongside these technological advancements to ensure that this collaborative future is inclusive and equitable.

More than just a tool or a partner in creation, AI prompts us to reconsider what creativity means. We might ask: Is AI's creative capacity distinct from human creativity, or is it an extension of human creativity itself? As we collaborate with AI, we're not merely transferring our creative processes but are actively engaging in a dialogue that transforms both AI and human participants.

AI-driven creative collaborations herald a profound shift in our understanding of art and creativity. They compel us to rethink how we produce, perceive, and value creative artifacts. While AI doesn't replace the human touch, it magnifies it, enabling artists to exceed the limitations of their imagination. This synergy between human creativity and AI facilitates a creative explosion—a burst of originality that redefines what it means to create in the quantum age.

Thus, as we advance into this quantum AI era, we're not just witnessing the birth of new art forms and experiences, but we're also redefining the essence of creativity itself. The collaboration between human intuition and quantum-enhanced AI signals a transformative moment in art, where creators become pioneers of artistic frontiers, breaking free from traditional limits and venturing into a realm where imagination knows no bounds.

Chapter 17: Societal Implications of Quantum AI

The advent of Quantum AI promises to redefine societal structures, unleash unprecedented human potential, and catalyze profound transformations across diverse industries. As this nascent technology matures, we stand on the brink of an era where job markets could be radically reshaped, with automation elevating human capabilities to new heights rather than rendering them obsolete. Quantum AI could amplify intellect and intuition, crafting partnerships between humans and machines that are more synergistic than ever before. However, this revolutionary shift isn't without its complexities and challenges, especially concerning societal disparities. The disparity in access to Quantum AI resources could exacerbate existing inequalities, demanding inclusive policies and proactive measures. As we navigate this quantum frontier, it's crucial to ensure that the transformative power of Quantum AI is harnessed responsibly, fostering a future where its benefits permeate through all strata of society. With thoughtful governance and ethical foresight, Quantum AI has the potential to act as a great equalizer, rather than a divider, shaping a future that embraces technological marvels while staying grounded in humane values.

Job Market Transformations

The fusion of quantum computing and artificial intelligence (AI) promises a seismic shift in the job market, reshaping roles across industries while ushering in new ones that we might not yet fully comprehend. Picture a world where tasks once considered solely within the human domain are performed with unprecedented efficiency and speed. This isn't a distant future; it's on the horizon, and its implications are profound.

Quantum AI's transformative capabilities lie in its ability to process vast amounts of data with unparalleled speed. For professionals immersed in data-intensive fields such as finance, healthcare, and logistics, this signifies a pivot from traditional data analysis techniques. The demand for roles specializing in quantum algorithms and quantum machine learning will skyrocket as industries seek to harness these potent tools. The conventional job descriptions will evolve; analysts may find themselves becoming "quantum analysts," tasked with deciphering insights from quantum-processed outputs.

There's an anticipation of roles that straddle the realms of AI and quantum computing, requiring a hybrid set of skills. The emerging workforce will need fluency in quantum mechanics, computer science, and data science. Educational institutions are already recalibrating their curricula to include quantum computing as a core component, recognizing its integral role in future job readiness. Courses designed to prepare students for these new cross-disciplinary demands will populate university classrooms, evening workshops, and online platforms, cultivating a new breed of professionals schooled in the intricacies of quantum AI.

Transitioning from traditional roles to those in the quantum AI domain isn't simply a matter of learning new technical skills; it also involves honing soft skills. As automation and intelligent systems take over routine tasks, human creativity, problem-solving, and critical thinking will become even more valuable. The workforce will need to adapt, finding new ways to collaborate with machines and use quantum insights to drive innovation.

The repercussions of this technological amalgamation stretch into the manufacturing industry, where mechanized processes are reinvented with quantum precision. A factory manager, once tasked with optimizing production schedules, could soon find themselves consulting a quantum AI system for real-time adjustment recommendations to enhance efficiency and reduce waste. Quantum AI will redefine how production and supply chains operate, necessitating a workforce adept in managing these evolved processes.

Let's not forget the potential for redistributing the workforce geographically and economically. As quantum AI expands capabilities, industries may become less centralized. Remote sensors powered by quantum technologies could allow workers in rural or economically disadvantaged areas to participate in global markets, thereby decentralizing opportunities and fueling local economies. The landscape of job availability could become more equitable, offering opportunities where there were few.

However, with transformation comes disruption. Some jobs, particularly those that are repetitive and data-heavy, might face redundancy. Yet, history shows that innovation often sparks new job sectors. Consider the rise of the tech industry itself, which burgeoned from the revolution of personal computing and the internet, creating jobs that were unthinkable just decades ago. Quantum AI is poised to do the same, though perhaps with a broader scope and a deeper impact.

The psychological impact on the workforce can't be discounted. As we stand on the cusp of profound technological change, job insecurity may rise amid uncertainties about the future job landscape. Workers will need support to transition into new roles, pointing to a growing need for programs centered on upskilling and reskilling. Organizations should prepare to invest in such programs, recognizing that a well-supported workforce is crucial for adapting to quantum AI's advances.

This evolution also challenges traditional notions of job ownership and expertise. In many industries, long-held expertise might need to be re-evaluated and redefined in the light of new technologies. Industries will require "life-long learners," professionals who continuously adapt to new tools and methodologies brought about by quantum AI advancements.

Corporate structures may also morph to accommodate these changes. Hierarchical models might give way to more fluid, adaptable structures that emphasize interdisciplinary collaboration and innovation. Teams composed of professionals across quantum computing, AI, and domain-specific experts will spearhead initiatives that harness these technologies for groundbreaking solutions.

In navigating these changes, it is imperative to consider the broader societal implications. Policymakers must address the potential for job displacement, ensuring safeguards are in place for those who may be adversely affected. Social safety nets and comprehensive policies aimed at workforce transition will be crucial in maintaining social stability and equity.

The potential of quantum AI to transform the job market is vast, and its promise is boundless. As the interaction of quantum and AI redefines what is possible, it invites us into a future where the human-machine partnership reaches new heights. Preparing for this future requires foresight, adaptability, and a commitment to leveraging these technologies to enhance and enrich our lives for the better.

Enhancing Human Capabilities

The convergence of quantum computing and artificial intelligence (AI) presents exciting possibilities for enhancing human capabilities.

By leveraging the unique strengths of these emerging technologies, we can augment cognitive functions, overcome neurological limitations, and fundamentally reshape how we interact with machines and each other.

Imagine a world where quantum-enhanced AI systems can process vast datasets with unprecedented speed and accuracy. Such systems could provide insights previously unattainable with classical computing. By identifying patterns and correlations at a probabilistic level, quantum AI can help individuals make better decisions, solve complex problems, and optimize processes in real-time. For instance, in fields like scientific research, quantum AI could enable researchers to explore new hypotheses and generate novel insights far beyond the scope of current methodologies.

The potential benefits extend beyond problem-solving. Quantum AI could revolutionize personal productivity by offering personalized tools that adapt to individual workflows and learning styles. Tailoring tasks and information to suit personal needs, quantum AI could significantly enhance individual efficiency and creativity. Professionals in any field could transform raw data into actionable knowledge, ultimately driving innovation.

Quantum AI may also play a pivotal role in expanding human knowledge. By simulating complex scenarios and environments, it can expedite discovery in areas such as space exploration and biomedical research. Consider the rapid development of highly accurate virtual models of biological systems, which could accelerate drug discovery and lead to breakthroughs in treating chronic diseases. These advancements could dramatically increase life expectancy and quality of life, representing unparalleled enhancements in human capability.

Moreover, focusing on accessibility, quantum AI has the potential to empower individuals with disabilities. Imagine AI-driven devices capable of interpreting and translating human experiences, offering new ways to interact with the world for those with sensory impairments. Beyond compensating for physical or cognitive challenges, these technologies could enrich lives by providing tailored educational tools, enhancing communication, and offering new opportunities for growth and expression.

The educational sector also stands to benefit from these advancements. By employing quantum AI, educators can develop more personalized and adaptive learning environments, catering to diverse learning styles and paces. Students could receive customized learning paths and experiences, unlocking their full potential irrespective of existing limitations. Such systems might even predict and address learning obstacles before they arise, encouraging more equitable educational opportunities worldwide.

Another area of promise lies in enhancing emotional intelligence and interpersonal skills. Quantum AI systems could assist in interpreting nuanced human emotions and behaviors, forming deep connections and understanding in both personal and professional relationships. Through these interactions, we could cultivate empathy and improve societal cooperation, laying the groundwork for more harmonious communities.

As we stand on the precipice of these transformative capabilities, ethical considerations and responsible stewardship become paramount. It's crucial to develop and implement these technologies with a focus on inclusivity and fairness, ensuring that enhancements to human capabilities do not widen existing societal disparities. Initiatives must be put in place to ensure equitable access to quantum AI technologies, empowering all demographics to benefit from its advancements.

Quantum AI promises a radical shift in human-machine collaboration, redefining the boundaries of what is achievable. As these tools become increasingly embedded in daily life, they will catalyze profound changes across industries and personal domains. However, the challenge lies in steering this development toward a future where quantum AI serves as a catalyst for human flourishing, elevating potentials and fostering an enriched, connected world.

In the face of both opportunity and uncertainty, let us strive to harness the power of quantum AI to uplift humanity, breaking the chains of limitation and opening the door to uncharted possibilities. By doing so, we'll set forth on a journey toward a future defined not by the mere augmentation of human capability but by the creation of new paradigms for human existence itself.

Societal Disparities and Quantum AI

The dawn of quantum AI promises a renaissance in technology and intelligence, yet it holds the potential to exacerbate existing societal disparities. As advancements in quantum computing and AI steadily progress, they could widen the gap between those with access to cutting-edge technology and those without. This digital divide may manifest both between countries and communities within them, leading to unequal opportunities in education, healthcare, and economic advancement. Bridging this chasm requires strategic foresight and ethical consideration.

Education stands at the forefront of this divide. As quantum AI technologies find their way into classrooms, they offer extraordinary opportunities for personalized learning and curriculum enhancements. However, schools and educational institutions that lack funding or connection to technological hubs may find themselves unable to leverage these innovations. If left unchecked, this disparity can result in a generation educated at differing technological standards — some equipped with quantum AI-powered tools, guiding them towards future-centric careers, while others remain versed in obsolete methodologies.

In healthcare, quantum AI holds the promise of personalized medicine and faster, more accurate diagnosis. Institutions with the infrastructure to implement these technologies could drastically improve patient outcomes. Yet, regions without the technological means may lag further behind in healthcare standards. The contrast between regions with advanced quantum AI medical solutions and those without could become stark, leading to significant inequalities in health outcomes and life expectancies across different populations.

The economic implications are notable as well. Quantum AI's integration into industries like finance, manufacturing, and logistics offers substantial improvements in efficiency and productivity. Organizations able to adopt these technologies can gain significant competitive advantages, driving economic growth. However, small businesses or companies in less developed regions may struggle to compete or access the requisite technology, leading to an economic imbalance that favors established players and technological elites.

Moreover, the ability of nations to participate in the quantum AI revolution will likely influence their geopolitical power. Countries investing heavily in quantum and AI research may emerge as global leaders, influencing policy and economic standards worldwide. Contrastingly, nations with limited access to these technologies might find their global influence diminished. This divide could lead to a new class of digital colonialism, where technological power plays a more prominent role in shaping global hierarchies.

Thus, it becomes imperative to address these disparities through inclusive policies and equitable technology dissemination strategies. Governments, industry leaders, and academia need to collaborate to ensure that the benefits of quantum AI are not just concentrated among the privileged classes but extend to marginalized and developing communities as well. This necessity calls for international cooperation to establish frameworks and guidelines that prioritize equitable access to quantum resources, investments in infrastructure, and educational outreach.

Investment in education and training programs that focus on quantum computing could be a key strategy to mitigate disparities. By creating opportunities for underrepresented groups to enter quantum AI fields, the workforce can become more diverse and inclusive. Additionally, fostering open-source and accessible quantum AI platforms can democratize access, enabling a broader range of people and organizations to engage with and innovate within this new arena.

Finally, awareness and public discourse must accompany these technological strides to ensure society navigates the ethical landscape of quantum AI equitably. Public understanding and dialogue about the capabilities and potential risks of quantum AI are vital. Without widespread education and open discussions about this technology, unintended consequences and fears could hinder its adoption and potential to uplift diverse communities.

Quantum AI possesses the transformative power to redefine society, industry, and day-to-day life. Its integration presents both phenomenal opportunities and risks of deepening societal disparities. By proactively addressing challenges related to equitable access and integration, we can foster a world where quantum AI benefits everyone, rather than perpetuating existing divides.

Ensuring an egalitarian deployment of quantum AI aligns with broader goals of social justice and equality. By narrowing the digital divide, we harness the true potential of these technologies to empower global communities, uplift economies, and enhance human capabilities across the spectrum of society. Embracing innovation while vigilantly counteracting its polarizing tendencies will shape a future where technology serves as a bridge rather than a barrier.

Chapter 18: Overcoming Technical Challenges

In the rapidly evolving landscape of quantum computing and artificial intelligence, one of the main challenges lies in overcoming technical hurdles that impede transformative progress. Building scalable quantum systems is crucial, as the potential for quantum computing to revolutionize industries depends on its ability to handle increasingly complex algorithms and massive datasets. Yet, scalability introduces the problem of quantum noise, requiring sophisticated error correction techniques that are vital for ensuring reliable calculations. Alongside these complexities in quantum mechanics, AI optimization techniques must adapt to accommodate quantum architectures, creating a harmonious interaction between classical and quantum calculations. By addressing these challenges head-on, we pave the way for pioneering solutions that not only promise to enhance computational power but also open doors to unprecedented technological advancements. The journey of overcoming these challenges is not merely about engineering prowess; it is also about inspiring a future where the convergence of quantum computing and AI leads to breakthroughs that we are only beginning to imagine. Through continuous innovation and collaboration, we will unlock new potentialities, fundamentally altering the technological and societal landscape.

Scalability of Quantum Systems

In the quest to harness the transformative power of quantum computing, scalability stands as a pivotal challenge. The unparalleled potential of quantum systems arises from their ability to process vast and complex calculations at incredible speeds. However, the journey from a few dozen qubits in laboratory environments to millions necessary for practical applications presents a series of formidable barriers. As researchers fan out worldwide to overcome these obstacles, they unveil the intricacies and constraints inherent in scaling quantum systems.

Currently, most quantum systems operate with tens of qubits at the experimental level. These quantum bits, or qubits, are the foundational units of quantum information. They differ significantly from classical bits, as they can exist in multiple states simultaneously due to a phenomenon called superposition. Moreover, qubits can be entangled, enabling unprecedented interconnectedness and complexity in computing. While these properties enhance computational power, they also introduce challenges in maintaining coherence over numerous qubits, especially as systems scale. The task becomes increasingly complex when trying to manage the delicate state of qubits for extended calculations.

One major hurdle in scalability is maintaining qubit fidelity. Quantum systems are notoriously susceptible to errors caused by decoherence and noise, primarily due to environmental interactions. In essence, qubits are incredibly sensitive, and their state can deteriorate in very short periods, often measured in microseconds. As quantum systems grow, ensuring that each qubit maintains its intended state without interference becomes exponentially more difficult. Error rates inevitably rise, complicating the goal of achieving truly reliable quantum computations on a large scale.

Another significant challenge is connectivity. In a small-scale quantum system, individual qubits can be directly entangled with one another. But as the number of qubits increases, efficiently managing their connections becomes a daunting task. The complexity of the system grows, demanding sophisticated architectures that can handle the increased interaction patterns without collapsing under their own weight. Developing scalable architectures that can accommodate thousands or even millions of qubits entails meticulous attention to the limitations of current technology and innovative foresight to leap over these hurdles.

To address these challenges, researchers are exploring multiple avenues. One approach is the development of quantum error correction codes, designed to protect information in qubits by encoding it across multiple qubits together. While this theoretically allows quantum systems to operate fault-tolerantly, implementing these codes requires a significant overhead of auxiliary qubits. Thus, achieving error correction in a scalable manner without significantly expanding the system dimensions remains an ongoing area of intense research and development.

Researchers are also investigating hybrid systems that combine classical and quantum computing elements. These platforms aim to leverage the best of both worlds, using classical computers to manage simpler tasks or certain error corrections while reserving quantum processors for exceptionally complex computations. However, an additional layer of complexity emerges from ensuring seamless integration between these different computing paradigms, particularly as the quantum elements scale.

Material science innovations play a crucial role in the quest for scalability. The drive to discover and develop new materials that can effectively serve as qubits with high stability and coherence time is active. Current qubit implementations, such as superconducting circuits, trapped ions, and even topological qubits, possess their own sets of pros and cons. Continuous advances in material science could unlock new forms of qubits that can improve the overall scalability of quantum systems.

Moreover, advancements in machine learning and artificial intelligence hold promise for addressing scaling challenges in quantum systems. These technologies can optimize numerous processes involved in quantum computations, such as error correction, qubit connectivity optimization, and resource management. By examining vast data sets and identifying patterns beyond human capacity, AI can provide insights that catalyze innovations in scalable quantum systems, driving them toward more robust and versatile applications.

The quest for scalable quantum systems is a complex interplay of physics, engineering, computer science, and beyond. Collaboration across these disciplines is paramount, suggesting avenues where interdisciplinary teams can forge new paths in the search for solutions. As research progresses, the community may need to embrace novel approaches and methodologies, some of which might yet be unexplored or under development. Scalability challenges demand not only technological and scientific breakthroughs but also strategic collaboration and cooperation across borders and industries.

The journey toward scalable quantum systems reflects a broader narrative—a dynamic dialogue between nature's fundamental laws and human ingenuity. Overcoming the technical challenges of scalability is not merely about surmounting obstacles but also about redefining the very paradigms of computational capability. As we stand on the edge of this quantum frontier, the pursuit of scalability serves as both a barrier and a beacon, driving the innovation that will unlock unparalleled possibilities in the realms of computing, technology, and the collective future of intelligence.

Error Correction in Quantum Computing

As we delve into the labyrinth of quantum computing, one of the most formidable challenges we encounter is error correction. Unlike classical computers, which deal in the binary absolutes of 0s and 1s, quantum computers revel in the ethereal complexity of qubits. These qubits, while powerful with their ability to exist in superposition and become entangled, are also notoriously prone to errors. These inaccuracies stem from their sensitivity to external disturbances—a whisper of noise or the slightest temperature fluctuation can disrupt them, challenging the reliability and stability of quantum computations.

Implementing a robust error correction system in quantum computers feels akin to building a ship that must remain seaworthy amidst the tempestuous seas of quantum chaos. Classical error correction methods, which work efficiently with binary data, falter when applied to quantum data. This is because qubits can't be simply examined without altering or destroying the information contained within them. Hence, novel strategies must be devised to correct errors without directly measuring and thereby disrupting quantum states.

The foundation of quantum error correction lies in redundancy and the development of error-correcting codes specifically for qubits. This involves encoding a single logical qubit into multiple physical qubits, often through schemes like the Shor Code or the Surface Code. These codes endeavor to protect quantum information by spreading it over many qubits so that even if one or more qubits experience errors, the original data can still be salvaged.

One might wonder, how is it possible to detect an error without directly observing the qubits? Quantum error correction exploits the genius of measuring auxiliary qubits—called syndrome qubits—that are entangled with the logical qubits. When an error occurs, the state of these syndrome qubits changes, flagging a possible error without compromising the data stored in the logical qubits themselves. Hence, a quantum computer can autonomously identify which errors have occurred and apply the necessary corrections, much like how classical computers use parity bits.

Despite these innovations, the real-world application of quantum error correction remains a Herculean task. The principal impediment is the sheer number of physical qubits required to realize a single logical qubit. Current estimates suggest that creating a fault-tolerant quantum computer will require thousands—if not millions—of physical qubits to implement effective error correction. As such, while the theory is sound and promising, the path to practical quantum error correction necessitates breakthroughs in qubit technology, coherence time, and miniaturization.

Continual advancements in error correction methods are critical as they directly influence the scalability and viability of quantum computing technologies. Researchers are tirelessly working on refining existing codes, testing new protocols, and fabricating qubits that are less susceptible to decoherence and operational errors. Innovations like topological qubits, which use braiding and anyons to encode information, hold promise as they inherently provide fault tolerance and reduce the overhead of error correction.

In addition to hardware advancements, software solutions and algorithmic improvements play a crucial role in addressing error correction challenges. Quantum algorithms are being designed to be more error-resistant, thereby reducing reliance on massive error correction redundancies. Hybrid quantum-classical algorithms provide another layer of protection by leveraging classical computation to supplement and verify quantum results.

The imperative of masterfully overcoming error correction is not just technical but also profoundly aspirational. Error correction is the linchpin on which the promise of quantum computing hinges. It offers a tantalizing bridge to a future where the processing power of quantum computers can solve problems that are currently intractable, spanning from fully functioning quantum simulations for material science to transformative solutions in AI.

The journey of error correction in quantum computing is not only marked by challenges but also filled with revolutionary opportunities. It exemplifies how human ingenuity continually meets and overcomes obstacles through creativity and perseverance. As research copes with the demands for more qubits and scalable systems, each step forward brings us closer to unlocking quantum computing's full potential, transforming industries and even the fabric of society in ways that are only beginning to be imagined.

Looking ahead, the quest for impeccable error correction may well redefine how we perceive computation itself. It encourages a profound reflection on the epistemological boundaries of scientific understanding and technological development. We stand at the cusp of a paradigm shift, where overcoming these errors is not merely a technical feat but a step towards a new reality—an empirical renaissance driven by quantum dynamics.

In conclusion, while the path to error-free quantum computing is fraught with hurdles, it is a journey worth embarking on. The knowledge and innovations we gain along the way will not only fortify the foundations of quantum computing but also foster a deeper comprehension of the quantum realm. Through persistent exploration and collaboration across disciplines, the dream of operational quantum systems will become a reality, illuminating a new era of computational dominance.

AI Optimization Techniques

As we delve further into the convergence of quantum computing and AI, it becomes essential to focus on optimizing AI algorithms to harness the extraordinary computational capabilities that quantum machines offer. This synergy carries the potential to surmount some of the most persistent technical challenges in AI, making optimization techniques paramount to achieving breakthroughs. The crux of AI optimization involves refining algorithms to be faster, more efficient, and capable of solving more complex problems than ever before.

Today's AI systems, heavily reliant on classical computing, often encounter bottlenecks in speed and efficiency, especially when handling massive datasets. Classical solutions struggle with high-dimensional problems, limiting their capabilities in fields such as natural language processing and complex simulations. Here, optimization techniques specific to AI aim to streamline processes, reduce error rates, and improve learning efficiency. These improvements become vital as developers strive to implement AI solutions capable of optimizing processes across various industries.

One foundational AI optimization technique revolves around parameter tuning. In the context of neural networks, for example, the practice of adjusting hyperparameters—such as learning rate, batch size, and network architecture—is crucial for performance enhancement. This process becomes even more intriguing when we consider quantum computing's potential impact on such optimization problems. Quantum annealing, an optimization approach tailored to find the global minimum of complex functions, offers a promising avenue. With its ability to manage multiple possibilities simultaneously through quantum superposition, quantum annealing could significantly accelerate solving NP-hard problems.

Furthermore, the integration of reinforcement learning with quantum computing showcases a fascinating area of exploration. Reinforcement learning, a method where agents learn by interacting with their environment to maximize some notion of cumulative reward, benefits tremendously from optimization. Quantum-enhanced reinforcement learning can capitalize on quantum processing's parallelism, exploring a vast array of potential policy strategies concurrently, thereby making optimization exponentially faster and more robust than classical methods.

Moving beyond reinforcement learning, we find that quantum-inspired algorithms are an optimizational revelation. These algorithms, inspired by quantum principles but designed to run on classical systems, offer unique optimization benefits. They work by leveraging quantum principles like superposition and entanglement, albeit in a theoretical rather than practical sense, to enhance classical algorithms. Examples include quantum-inspired versions of classic optimization algorithms like the Genetic Algorithm and Particle Swarm Optimization, which have shown remarkable improvements over their classical counterparts.

The path to AI optimization also embeds itself deeply in the realm of neural networks. Quantum neural networks (QNNs) present a cutting-edge approach to reducing computational complexity. Optimizing these networks involves configuring qubits to process information in novel ways that classical bits cannot, thus optimizing both learning processes and computational resources. As quantum technology matures, the optimization of neural networks via quantum entanglement and gate transformations may redefine the efficiency of training processes and the dimensionality of the data handled.

Another facet of AI optimization lies in machine learning model compression techniques. Methods such as pruning, quantization, and knowledge distillation aim to shrink model sizes without compromising performance. These techniques become pivotal as the size and complexity of datasets balloon. Implementing these methods in combination with quantum machines leads to reduced latency and lower energy consumption, creating more sustainable solutions in AI deployment. Remarkably, the reduction in model size can have far-reaching implications for deploying AI in resource-constrained environments, such as edge devices.

In addition to these methods, adversarial training emerges as a formidable tool in AI optimization. By deliberately introducing slight perturbations during the training phase, adversarial training strengthens model robustness. When paired with quantum computing, this technique can expand a model's ability to withstand variability and noise in data, resulting in systems that are not only optimized for performance but also for reliability and security.

Efficient data preprocessing also sits at the heart of AI optimization techniques. Transforming raw data into a suitable format for processing can significantly enhance model performance. Techniques such as data normalization, dimensionality reduction, and feature selection are crucial. Quantum computing can contribute to preprocessing by using quantum algorithms that analyze vast datasets more rapidly and identify data patterns or features that classical methods might overlook.

At the broader intersection of these techniques, lies the potential for quantum AI platforms that are self-optimizing. Imagine systems that automatically adjust their parameters, architectures, and learning strategies in real-time, based on current performance and environmental feedback. These platforms represent the zenith of AI optimization: self-improving and adaptable systems capable of evolving to meet ever-increasing demands in real-time.

Ultimately, AI optimization techniques are a transformative force that can revolutionize how AI is deployed across various industries. The interplay of these techniques with quantum computing promises not just incremental improvements but paradigm shifts in computational power and efficiency. It opens doors to solving previously intractable problems, harnessing more profound insights from data, and enabling smarter, more capable AI systems. The future of AI optimization looks towards a horizon where quantum-enhanced techniques are not just supplemental but integral to the functioning and advancement of artificial intelligence.

Chapter 19: Current Research and Development

The frontier of quantum computing and AI is a whirlwind of innovation, brimming with possibilities that extend beyond conventional paradigms. In labs around the world, researchers are pushing the envelope, crafting algorithms that defy present limitations and harnessing the enigmatic potential of qubits to power intelligent systems. Major institutes like MIT and Google are at the vanguard, fostering environments where groundbreaking theories become tangible breakthroughs. Collaborations between tech giants and academic powerhouses have catalyzed a surge in development, fostering a community that thrives on shared knowledge and open innovation. The focus is not just on theoretical advancements but also on practical applications that can revolutionize industries—whether it's through accelerating drug discovery, optimizing logistics, or enhancing cybersecurity. These cutting-edge developments are setting the stage for a new era, one where the synergy of quantum and artificial intelligence redefines the fabric of technology, inspiring a future teeming with endless possibilities.

Leading Institutes and Researchers

In the rapidly evolving field of quantum computing and artificial intelligence, certain institutes and researchers have emerged as key players, propelling the discipline forward with groundbreaking contributions. These hubs of innovation are not just incubators of knowledge but are also pioneering collaborations that span continents, languages, and technologies.

The Massachusetts Institute of Technology (MIT) stands tall in the landscape of quantum research, with the MIT-IBM Watson AI Lab leading the charge. Their work focuses on understanding the fundamental limits of AI through quantum computing. By devising quantum algorithms that promise unprecedented computational power, they aim to redefine the boundaries of what can be computed. The collaboration between MIT and IBM illustrates a profound synergy, where academia and industry converge to tackle some of the most daunting challenges of our time.

Over in California, Stanford University, with its Center for Quantum Computation and Advanced Physics, continues to be at the forefront of theoretical and applicable quantum physics. Researchers here are making significant strides in developing error-correction strategies and scalable quantum architectures, which are critical for making quantum computers practical and efficient. Stanford's approach is unique in that it combines deep theoretical exploration with hands-on experimental work, thus bridging the gap between quantum theory and its real-world execution.

Across the pond, the University of Oxford and its Department of Computer Science have become synonymous with quantum innovation. Home to some of the brightest minds in the field, Oxford is diving deep into quantum algorithms for AI. Their focus has been on how to expand classical algorithms' capabilities using quantum principles, especially in machine learning and optimization problems. The work being done here doesn't just push theoretical boundaries but also promises to offer tangible enhancements in various AI applications.

The role of international cooperation cannot be overstressed, as seen with the European Union's Quantum Flagship initiative, which commits to a decade-long investment in quantum technologies. This ambitious project encompasses numerous research institutions and universities across Europe, working collectively to harness quantum computing's potential. By funding large-scale projects and fostering collaboration between different entities, the Quantum Flagship aims to propel Europe into a competitive global leader in quantum developments. Such partnerships have yielded significant outcomes, illustrating how shared knowledge and resources can accelerate scientific progress.

China has also entered the quantum arena with vigor and determination. Institutions like the University of Science and Technology of China (USTC) in Hefei lead Chinese efforts in quantum information science. USTC's work has been pivotal in demonstrating quantum supremacy with photonic systems, a milestone that highlights China's growing capabilities in this field. The Chinese government's substantial funding and policy support underscore the country's strategic emphasis on quantum technologies as a pillar for future growth.

Collaborations between industry giants and academic institutions are becoming increasingly prevalent. Companies like Google and Rigetti Computing have fostered strong partnerships with universities worldwide, enabling them to stay at the cutting edge of quantum research. These collaborations are vital, as they not only bring together theoretical insights and business acumen but also accelerate the translation of complex quantum ideas to market-ready solutions.

Furthermore, Canada's Institute for Quantum Computing at the University of Waterloo is another beacon of progress. Known for its multidisciplinary approach, the institute blends the expertise of scientists from physics, engineering, mathematics, and computer science. Their research focuses on foundational quantum mechanics, with a special interest in how quantum technologies can be integrated with AI to achieve breakthroughs in complex data processing and cryptography.

The ongoing research at these leading institutes would not be possible without the visionaries who lead them. Individuals such as John Preskill from Caltech, who coined the term "quantum supremacy," have guided research trajectories not with mere innovation but with a broader understanding of the ethical, societal, and practical implications of their work. Their influence extends beyond academic circles, informing policies and guiding international standards in quantum computing practices.

In this landscape of rapid transformation, many researchers operate under a shared guiding principle: the integration of ethical considerations with cutting-edge science. As AI and quantum computing continue to merge, the role of these researchers will involve not only pushing technological boundaries but also ensuring the responsible development of these powerful tools. This balance is especially crucial as quantum AI technologies start to enter sectors that directly impact human lives, such as healthcare, finance, and transportation.

The development of quantum AI is not a solitary endeavor. It requires the concerted efforts of researchers, policymakers, and industry leaders working together to overcome challenges and harness opportunities. Each step forward builds upon the legacies of pioneering work being done across the globe, creating a tapestry of innovation that's both locally grounded and internationally relevant. With these leaders at the helm, the future of quantum AI is brimming with promise, shaping a world where the impossible becomes possible.

Breakthroughs and Milestones

The landscape of quantum computing and artificial intelligence has witnessed unprecedented breakthroughs that continue to push the boundaries of what's possible. These advancements have not only marked significant milestones in specific fields but have also paved the path for future innovations, creating ripples across many industries. From fundamental discoveries in quantum mechanics to integrating AI's capabilities with quantum systems, the trajectory of these developments presents a fascinating tapestry of human ingenuity and technological progress.

Among the most remarkable breakthroughs is the creation and operationalization of quantum supremacy. This term, coined to mark the moment when a quantum computer can solve a problem that classical computers find intractable, was first realized in 2019 by Google. Their Sycamore processor was able to perform a computation in seconds that the world's fastest supercomputers would have taken millennia to accomplish. This event was not merely a technical accomplishment but a symbolic leap forward, validating the potential of quantum computing to revolutionize problem-solving strategies across multiple domains.

Parallel to these feats in hardware, advancements in quantum algorithms have been equally transformative. Shor's algorithm, for example, demonstrated the potential to factorize large numbers exponentially faster than any known classical algorithm, threatening the cryptographic schemes that secure digital communications today. Such developments have prompted a reevaluation of cryptographic protocols worldwide, sparking a flurry of research into quantum-resistant algorithms as nations and corporations prepare for a post-quantum world.

On the AI front, deep learning breakthroughs have significantly advanced autonomy and decision-making processes. The invention of transformer models, exemplified by attention mechanisms, has revolutionized how machines understand and generate human language. OpenAI's GPT models have garnered attention for their ability to produce human-like text, heralding a new era of AI applications in natural language processing. These innovations have profound implications for education, content creation, and communication technology.

The intersection of AI and quantum computing, often called Quantum AI, is a burgeoning area of research that holds promise for further breakthroughs. Quantum machine learning, which leverages quantum algorithms to process and analyze vast datasets, stands poised to transform fields like genomics, pharmaceuticals, and climate modeling. Imagine harnessing the power of quantum states to find patterns in complex data that were previously indecipherable. Such capabilities could accelerate the discovery of new drugs, optimize supply chains, and even model intricate weather systems with unprecedented accuracy.

Furthermore, pioneering efforts in hybrid quantum-classical systems are showing great potential. Bridging the robustness of classical computing with the unique capabilities of quantum systems, researchers have reported milestones in developing algorithms that promise exponential speed-ups for optimization problems. These systems provide a practical path forward, offering scalable solutions while quantum technology matures and overcomes existing technical challenges like error correction and qubit stability.

Collaboration remains a cornerstone of these advancements. International alliances among tech giants, academic institutions, and governments have been instrumental in accelerating research and development. For instance, the partnership between IBM, several universities, and research labs aims to democratize access to quantum technologies via cloud-based platforms. This collaborative approach ensures that innovations are robust, inclusive, and designed to address global challenges comprehensively.

These significant milestones are not only technical by nature but have profound socio-economic implications. Governments around the world are investing heavily in quantum AI research, recognizing its potential to bolster national security, transform healthcare, and drive economic growth. The European Union's Quantum Flagship and China's nationwide quantum initiatives represent strategic commitments to becoming leaders in the quantum race, signaling a shift in global tech-power dynamics.

In this rapidly evolving field, milestones are not static but continually unfolding as new discoveries build upon previous breakthroughs. Startups and established companies alike are pushing boundaries in quantum materials and AI integration, creating a dynamic industry landscape teeming with innovation and potential. As we look to the future, the remarkable strides made thus far provide a glimpse into a world where quantum AI might redefine the contours of technology, driving interdisciplinary advancements and forging new paradigms in how we understand and interact with the universe.

Ultimately, the journey of quantum AI breakthroughs is as much about the technology as it is about the visionaries and researchers who dare to chase possibilities beyond conventional limits. Their quest for knowledge and solutions continually refines our understanding of the universe's fundamental workings, inspiring the next generation of inventors and thinkers to dream even bigger. As we stand on the threshold of the quantum age, each milestone represents not just a destination reached but a new beginning, with limitless horizons beckoning.

Collaborations and Partnerships

In the rapidly evolving landscape of quantum computing and AI, collaborations and partnerships play a pivotal role, acting as catalysts for innovation and groundbreaking research. A synergy between institutions, industries, and governments can create a fertile ground for the cross-pollination of ideas that drive both fields forward. As quantum technologies and AI unveil new horizons, partnerships become the bedrock of an ecosystem that thrives on shared knowledge and resources. Such alliances not only bridge the gap between theory and application but also amplify the potential reach and impact of research initiatives.

Consider the collaborative endeavors between academia and industry. Universities often house the foundational research, a cauldron of experimentation and theoretical exploration. By partnering with tech companies, these institutions can translate their theoretical provenances into real-world applications. For instance, major tech firms often provide the robust computational power and datasets necessary for validating advanced quantum algorithms and AI models. This symbiotic relationship accelerates the pace at which these cutting-edge technologies are tested, iterated, and potentially deployed at scale.

Governmental bodies also emerge as crucial stakeholders in these collaborations. Acknowledging the transformative potential of quantum computing and AI, many governments globally are actively funding initiatives and fostering international collaborations. This can include direct investments, creating national strategies for quantum development, or facilitating international research projects. With the stakes so high, especially concerning national security or economic leadership, governments understand that nurturing talent and innovation clusters within their borders is imperative.

The intersection of quantum computing and AI has also birthed a new wave of international alliances. Global challenges such as climate change, cybersecurity, and healthcare demand solutions that transcend borders, making international cooperation not just beneficial, but necessary. Through consortia or dedicated research institutes, countries are pooling resources, sharing findings, and tackling issues that are too colossal for any single entity to resolve independently.

Moreover, partnerships are not restricted to just the technology and research sectors. Cross-industry collaborations are equally crucial in translating quantum and AI advancements into everyday solutions. Companies from diverse sectors, including healthcare, finance, automotive, and agriculture, are increasingly partnering with quantum and AI startups. They are looking to leverage these technologies for improvements such as optimizing supply chains, enhancing predictive analytics, and bolstering data security. These collaborations enable different industries to access cutting-edge technologies, often democratizing access to advancements that would otherwise be beyond reach.

The rise of startup ecosystems centered around quantum computing and AI demonstrates another facet of collaboration. Startups often work at the cutting edge, offering agility and innovative approaches that larger organizations might find challenging to implement quickly. By forming partnerships with larger, established corporations, startups gain access to capital, mentorship, and market access, while larger firms benefit from fresh ideas and innovative solutions. This dynamic creates a vibrant tech ecosystem where bold ideas can flourish.

Non-profit organizations and think tanks also contribute significantly to this collaborative network. These entities often act as neutral grounds for dialogue, providing platforms for multi-stakeholder partnerships. By fostering discussions and documenting best practices, they contribute to setting standards, ethical guidelines, and public policy that govern the responsible development of quantum AI technologies.

The importance of collaborations and partnerships in research and development is not just about resource-sharing; it's about nurturing an environment where collective problem-solving can thrive. As complex as they are, the challenges that quantum computing and AI aim to tackle require a concerted effort that leverages diverse expertise and perspectives.

However, navigating these partnerships isn't without its challenges. Differences in organizational cultures, priorities, and timelines can sometimes hinder the progress of collaborative endeavors. Ensuring clear communication and aligning objectives are vital to overcoming these hurdles. Mechanisms for data sharing, joint intellectual property rights, and transparency protocols need to be well defined to foster trust and collaboration.

Looking to the future, fostering a culture of open innovation and collaboration will likely continue to be a strategic priority among leaders in technology and academia. The expectation is that as quantum computing and AI technologies mature, the networks of collaboration will become more intricate and impactful, reaching into more sectors and domains.

As part of this ongoing journey, educational partnerships are equally significant. Joint programs among universities, tech giants, and research institutions are setting the stage for training the next generation of scientists and engineers who will spearhead future innovations in these fields. Such educational alliances ensure that the workforce is well-prepared to handle the challenges of tomorrow and inject new vigor into the development of quantum and AI technologies.

Overall, the trajectory of quantum computing and AI underlines an era marked by unprecedented collaboration. By effectively engaging in partnerships across sectors and nations, the community not only propels technological advances but also cultivates an ecosystem poised to address some of the world's most pressing issues. The holistic integration of efforts across the board paints a promising picture where collaboration transforms potential into reality, creating a landscape where quantum computing and AI don't just exist, but thrive, bringing forth a future replete with possibilities.

Chapter 20: Investing in Quantum AI

As we stand at the cusp of a technological revolution, the prospect of investing in Quantum AI offers a unique blend of challenges and opportunities that are hard to parallel. Emerging market trends point toward a future where the convergence of quantum computing and artificial intelligence reshapes industries, fostering innovations that were once deemed impossible. Investors are increasingly eyeing this domain, drawn by the magnetic potential for unprecedented breakthroughs in data processing, predictive analytics, and decision-making frameworks. Venture capital and funding dynamics in the Quantum AI space are evolving rapidly, underscored by a surge in strategic partnerships and collaborations pushing the envelope of what can be achieved. The economy of Quantum AI is not just about short-term gains; it's a long-term commitment to nurturing technologies that promise to revolutionize everything from healthcare to financial modeling. Those who boldly step into this market today stand to not only capitalize financially but also contribute significantly to the dawn of a new era in computing and AI. With calculated risks and visionary investments, Quantum AI is poised to become a cornerstone of future economic and technological landscapes.

Market Trends and Opportunities

The quantum AI realm is not just on the brink of transformation; it is, in fact, reshaping entire industries and society as we know it. As we delve into the market trends and opportunities within this emerging field, it becomes evident that the potential for growth is unprecedented. Companies large and small are racing to harness the power of quantum computing combined with AI, setting the stage for a new wave of technological innovation.

One of the most significant trends in the quantum AI market is the rise of strategic partnerships between established tech giants and nimble startups. These collaborations are accelerating advancements by combining vast resources and revolutionary ideas. Recent years have witnessed a proliferation of these alliances, each aiming to conquer specific challenges or market niches. Companies like IBM, Google, and Microsoft are already at the forefront, developing quantum computing platforms while investing heavily in AI research.

Another compelling trend is the diversification of industries seeking to adopt quantum AI technologies. Initially, interest emerged primarily from sectors like finance and pharmaceuticals, but now we see automotive, logistics, and even creative industries exploring quantum solutions. The potential to revolutionize complex calculations, optimize logistical networks, and enhance product designs offers irresistible advantages for these industries. Startups dedicated to niche applications are becoming key players, offering specialized quantum algorithms tailored to industry-specific needs.

Investment in quantum AI is not just growing; it's exploding. Venture capital firms are pouring unprecedented amounts of funding into startups working at the intersection of quantum computing and artificial intelligence. According to recent reports, venture capital investments in quantum technology have surged by over 100 percent year-on-year. This enthusiasm is not just hype. It reflects a deep belief in the transformative potential of quantum AI to solve complex problems that traditional computing couldn't previously address.

With venture capitalists focusing on long-term gains, funding opportunities are expanding beyond seed and Series A rounds, covering later stages of startup financing. This marks a significant shift in the investment landscape, signaling a maturation process for quantum AI ventures. Investors, aware of the initial high capital requirements and long gestation periods, recognize that patience will be rewarded as quantum technology becomes more accessible and integrated into mainstream applications.

The geographical distribution of quantum AI development is also worth noting. While North America and Europe have been leading the charge, Asia is quickly catching up. Countries like China and Japan are making considerable investments in both infrastructure and talent to establish themselves as hubs for quantum AI research. Government support, in the form of grants and public-private partnerships, is catalyzing this growth, fueling a competitive arena where breakthroughs are no longer confined to traditional centers of tech innovation.

As we assess future opportunities, the potential integration of quantum AI in cloud services seems particularly promising. Cloud computing providers are starting to offer quantum computing as a service (QCaaS), making the immense computational power of quantum computers accessible to businesses without requiring them to maintain expensive infrastructure. This could democratize access to quantum AI tools, allowing even small enterprises to leverage these technologies to enhance their operations.

Moreover, the emergence of new business models centered on quantum AI is paving the way for fascinating opportunities. Quantum AI consultancy is an emerging field where experts assist companies in transitioning from classical computing models to quantum-enhanced frameworks. Additionally, the concept of "quantum AI as a service" is gaining traction, where companies can utilize ready-made quantum AI solutions to tackle specific problems, much like AI and ML models are used today.

As quantum AI technologies advance, ethical considerations and societal impacts will inevitably arise. Thus, another burgeoning opportunity lies in the development of frameworks and policies to ensure the responsible use of quantum AI. This includes addressing potential biases in algorithms, enhancing transparency, and safeguarding data privacy. Companies that can offer ethical guidance and robust policy solutions will play vital roles in shaping how quantum AI integrates into society.

Long-term market opportunities are abundant as strides in quantum AI lead to novel applications that once seemed like science fiction. Consider the potential implications in fields like material science, where quantum-enhanced models can predict molecular behaviors with unparalleled precision. In agriculture, quantum AI could revolutionize crop management through enhanced modeling of environmental variables. Innovative applications like these are just scratching the surface of what might be possible.

Moreover, as quantum AI ecosystems continue to develop, emerging infrastructure requirements will create additional market opportunities. From specialized hardware fabrication to advanced cooling systems and quantum-safe cybersecurity solutions, ancillary industries stand to benefit from the increased demand created by quantum AI deployments. This intricate web of opportunities will create a diverse marketplace with interconnected industries working in tandem to sustain growth.

Education and workforce development also present intriguing opportunities. As the demand for knowledgeable professionals in quantum AI grows, educational institutions and online platforms have a chance to play crucial roles in training the next generation of quantum AI specialists. Developing curricula that integrate quantum computing, AI, and interdisciplinary collaboration will be essential to prepare individuals for the rapidly evolving job market.

In summary, the quantum AI market is a fertile ground for innovation, investment, and societal advancement. The trends of strategic partnerships, diversified industry applications, burgeoning investment landscapes, and global expansion all signal a time of great change and promise. It's not just about solving today's problems but envisioning and creating the technology landscape of the future. For those who seize the opportunities, the promise of quantum AI is as vast as the boundless possibilities it unveils.

Venture Capital and Funding Dynamics

Investing in the field of quantum AI is akin to navigating a vast, uncharted ocean. The financial dynamics driving this domain are as intriguing as the technologies themselves. Venture capital, a cornerstone of innovation funding, plays a crucial role in providing the resources necessary to propel quantum AI from theoretical exploration to practical application.

At the heart of these dynamics lies an intricate tapestry of investment motivation. Quantum AI brings with it promises of extraordinary capabilities: solving complex problems unsolvable by classical computers, revolutionizing optimization processes, and drastically enhancing machine learning models. Venture capitalists are particularly enticed by these prospects, seeking opportunities for disruptive innovation that can redefine industry standards and open new markets.

Quantum AI's potential is not merely theoretical. Prominent success stories, though still emerging, act as beacon calls to investors. Companies at the forefront of this field, like Xanadu and Rigetti, have demonstrated tangible progress, drawing significant investment rounds. Their success highlights the appetite for early-stage companies that can offer a uniquely competitive edge in the emerging quantum landscape.

However, the excitement is tempered by the inherent risks associated with quantum AI investments. Unlike traditional tech ventures, quantum AI projects face profound scientific and engineering challenges. The pathway from a promising algorithm to a marketable product can be fraught with uncertainties. Investors must therefore possess not just deep pockets, but a robust understanding of the science and a tolerance for long gestation periods. The promise of high returns justifies these risks for many, but it's not a journey for the faint-hearted.

The evolving venture capital landscape in quantum AI is characterized by a shift towards strategic investments. Instead of merely funding startups, many venture firms are forming partnerships with research institutions and established tech companies to foster a more collaborative ecosystem. These alliances are essential for navigating the multifaceted challenges of quantum computing and AI development, from hardware constraints to algorithm optimization.

Moreover, government funding and initiatives play a pivotal role in this landscape. Recognizing the strategic importance of quantum AI, governments around the world have launched significant funding programs. These initiatives aim to bolster research, infrastructure development, and industry collaboration. Consequently, public funding often complements venture capital, creating unique opportunities for leveraging resources across sectors.

The competitive nature of venture capital in quantum AI also can't be understated. As quantum technology advances, the race to acquire top talent intensifies. Investors are keenly aware that human capital is as critical as financial capital. Ensuring access to leading quantum and AI experts is becoming a vital component of investment strategies, driving some venture firms to support educational programs and research initiatives directly.

Another dimension of funding dynamics in quantum AI involves the geographical distribution of investments. Silicon Valley, long hailed as the epicenter of technological innovation, still dominates, but regions like Europe and Asia are catching up. Government policies, cultural predispositions towards risk, and available infrastructure significantly influence where investors place their bets. This global diversification is accelerating the worldwide push towards quantum supremacy and AI integration, fostering a competitive yet fertile ground for innovation.

The financial models deployed by venture capitalists in quantum AI are adapting to this unique sector. Traditional equity investments are being supplemented with creative financial instruments, such as convertible notes and SAFE agreements, which reflect the high-risk profiles and long development timelines typical of quantum AI projects. This flexibility allows startups and investors to forge mutually beneficial relationships while managing the uncertainties inherent in cutting-edge technology.

Moreover, the influx of capital into quantum AI is catalyzing the development of specialized funds dedicated solely to this niche. These funds bring together investors who are not only financially invested but are also intellectually engaged with the field. This alignment of interests facilitates a more coherent vision for the evolution of quantum AI technologies.

As the chapter on venture capital and funding dynamics unfolds, it becomes increasingly clear that investing in quantum AI is not just about chasing the next big tech wave; it's about shaping the very future of technology. The financial landscapes, though complex, are driven by passions for discovery, innovation, and the potential to fundamentally transform how we understand and interact with the world.

Ultimately, the venture capital and funding dynamics in quantum AI reflect a grand tapestry of collaboration, competition, and aspiration. As investors navigate this intricate but rewarding domain, they serve as pivotal enablers of progress in what may very well be one of the most transformative technological advances of our era. Their role in fueling this revolution cannot be overstated, as they take calculated risks to forge pathways into the unknown, where quantum and artificial intelligence promise to redefine the boundaries of possibility.

The Economy of Quantum AI

The burgeoning field of Quantum AI stands at the intersection of two transformative technological advancements: quantum computing and artificial intelligence. As these technologies merge, they are poised to redefine entire industries and disrupt traditional economic structures. Understanding the economic implications of Quantum AI is crucial for investors, policymakers, and businesses aiming to leverage its potential. The market is ripe with opportunities, yet fraught with uncertainties that mirror the quantum states it seeks to manipulate.

Firstly, the potential economic impact of Quantum AI is immense. Just as classical AI has driven efficiency and innovation across sectors, Quantum AI promises to do the same, but on a much larger scale. Financial services, for instance, are on the cusp of a quantum revolution, as algorithms run on quantum computers could process vast amounts of financial data with unprecedented speed and accuracy. This ability could transform risk analysis and portfolio management, directly influencing market dynamics and economic strategies.

Moreover, the advent of Quantum AI could spur significant advancements in material sciences and pharmaceuticals. By simulating molecular interactions at a quantum level, new materials and drugs could be developed far more rapidly than is currently possible. This capability not only accelerates innovation but also reduces costs, thereby potentially lowering prices and making products more accessible—a boon for both consumers and the broader economy.

Investors are naturally drawn to the disruptive potential of Quantum AI. Venture capitalists and tech giants are pouring resources into startups that are developing quantum technologies and applications. However, investing in this nascent field requires an understanding of its unique challenges—chief among them, the technological uncertainty. Unlike classical computing, the development of quantum systems is still in its infancy, requiring a long-term investment horizon and a tolerance for volatility.

The economic landscape of Quantum AI is further shaped by the global race for quantum supremacy. Countries like the United States, China, and members of the European Union are investing heavily in quantum research, creating both opportunities and competitive pressures. Policies aimed at fostering innovation and protecting intellectual property will play a critical role in determining which regions lead the quantum age. The implications for the workforce are also considerable, as new skills and training programs will be essential to support quantum technological developments.

Meanwhile, the public sector has an indispensable role to play. Government funding and incentives could accelerate research, development, and deployment of Quantum AI technologies. Public-private partnerships might emerge as a critical strategy to share the risks and rewards of this transformative field. In an era where national security and economic growth are intricately linked to technological prowess, Quantum AI may become a linchpin of national economic strategies.

One cannot ignore the ethical and societal dimensions while mapping out the economy of Quantum AI. The wealth it generates will need careful management to avoid exacerbating social inequalities. If only a few corporations or nations control the quantum computing resources, we could see a deepening of the digital divide. Strategies for equitable distribution of quantum-derived wealth could become a field of study in itself, emphasizing the importance of inclusive policy-making.

The race to Quantum AI does not merely signify technological advancement but heralds an economic shift comparable to the digital revolution of the late 20th century. The creation of entirely new industries is a plausible outcome, alongside the disruption and evolution of existing ones. For instance, cybersecurity industries will be transformed when Quantum AI renders many current encryption methods obsolete, prompting an entire reevaluation of digital security frameworks.

The future economic landscape shaped by Quantum AI will hinge significantly on its ability to address technical limitations and scalability issues. Investments in infrastructure, talent, and research are crucial to overcoming these hurdles. As quantum technology matures, creating a robust ecosystem involving academia, startups, and established companies will be vital. The economy of Quantum AI is poised to foster a multidisciplinary collaboration like never before.

In conclusion, Quantum AI operates at the frontier of technology and economics, promising profound impacts on how industries function and compete. The fusion of quantum computing's immense processing power with AI's capabilities heralds a unique opportunity to drive unparalleled economic growth and innovation. However, with these opportunities come significant risks, requiring strategic foresight and cooperative global governance to harness Quantum AI's full potential responsibly.

Chapter 21: Quantum AI Policy and Regulation

As quantum AI emerges as a formidable force poised to reshape numerous sectors, its rapid evolution necessitates a robust policy and regulatory framework that both fosters innovation and ensures ethical adherence. The journey towards a comprehensive global regulatory structure involves balancing the acceleration of technological advancements with essential oversight to protect societal values and privacy. Policymakers are challenged to craft regulations that not only promote international cooperation and establish universal standards but also respect the diverse technological landscapes and socio-economic needs of different regions. By understanding the intricate dynamics of quantum AI, we can lay the groundwork for policies that facilitate transformative growth while mitigating risks. Strategic dialogue among nations is crucial, offering a collaborative platform to navigate the complexities of governance in this unprecedented technological era. The horizon holds immense potential as we seek the sweet spot between control and creativity, ensuring that the immense capabilities of quantum AI are harnessed responsibly and equitably for the benefit of all.

Global Regulatory Frameworks

In the realm of Quantum AI, the regulatory landscape is akin to a vast, uncharted territory. With the accelerated pace of technological advancements, a global regulatory framework must be established to ensure that the evolution of Quantum AI harmonizes with societal values and ethical norms. As Quantum AI combines the complexities of both quantum computing and artificial intelligence, it presents unique regulatory challenges that require a confluence of expertise from traditional computing, quantum physics, and AI ethics.

Currently, the regulation of Quantum AI exists as a patchwork of guidelines and standards, largely based on existing models for digital and AI technologies. However, the unique attributes of quantum computing—like superposition and entanglement—demand a tailored approach. These quantum features enable capabilities far beyond classical systems, necessitating the creation of bespoke regulatory measures. Such measures will not only need to account for technical intricacies but also consider the socio-economic ramifications associated with Quantum AI's broad applications.

International Cooperation is crucial in building an effective regulatory framework. Technology, by its very nature, transcends borders—prompting a need for global consensus on regulations. Bodies like the United Nations, the European Union, and international coalitions working under initiatives such as the Global Partnership on Artificial Intelligence can play pivotal roles in this global discourse. The goal should be to establish common standards while respecting the socio-political diversity among nations.

Consider, for instance, the General Data Protection Regulation (GDPR) in Europe, which serves as a benchmark for data privacy laws globally. A similar foundational framework can be envisioned for Quantum AI, ensuring data integrity and privacy while promoting innovation. Such regulations should be flexible, allowing room for evolution as the technology advances, yet robust enough to prevent misuse and protect public interest.

In crafting these frameworks, it's essential to address the inherent risks in Quantum AI, such as the potential for disrupting cryptographic systems. The cryptographic strength used to protect sensitive data today could be rendered ineffective by quantum computers' ability to solve complex mathematical problems exponentially faster than classical computers. Hence, regulatory bodies must consider a proactive approach in adopting quantum-resistant cryptographic methods.

Beyond technical concerns, ethical considerations also demand urgent attention. As Quantum AI systems become more autonomous, questions arise about accountability and transparency. Who controls these systems, and what happens when they misbehave? Ethical frameworks must be integrated into technical regulations to ensure that AI decisions reflect human values and can be audited by stakeholders. Regulatory principles should incorporate fairness, transparency, accountability, and traceability, ensuring that Quantum AI serves humanity equitably.

Another aspect requiring regulation is intellectual property (IP) rights. As advancements in Quantum AI spur significant innovations, there is a growing debate on how to balance IP protection with the collaborative nature of global research and development. An effective regulatory framework will need to address patent rights, considering both the rights of inventors and the benefits of open science.

The concept of *Regulatory Sandboxes* can serve as a forward-thinking approach, providing a controlled environment to test Quantum AI innovations under regulatory oversight. These sandboxes encourage experimentation, allowing innovators to operate temporarily outside existing regulations to test the feasibility and implications of new technologies. Through the implementation of such flexible regulatory mechanisms, regulators can gain insights into the technology's impacts before committing to rigid rules.

Furthermore, the role of public consultative processes cannot be overlooked. Inclusive dialogue with the public, academia, industry leaders, and policymakers about the ethos and potential of Quantum AI can ensure the technology aligns with societal values. Education and awareness play an essential part in contributing to informed discussions on regulation, fostering a public understanding that informs sensible policymaking.

It's vital that policymakers stay ahead of the curve, anticipating future challenges rather than reacting to crises. A forward-looking strategy involves fostering international coalitions that prioritize both technological advancement and ethical responsibility. Learning from historical instances where technology outpaced regulation could provide invaluable lessons in shaping a proactive and cohesive global regulatory stance.

In conclusion, formulating a global regulatory framework for Quantum AI is a formidable yet essential endeavor. While the task is complex, involving multiple stakeholders and balancing a myriad of competing interests, the potential benefits of a well-regulated Quantum AI ecosystem are profound. By ensuring a coherent and adaptable regulatory framework, we have the opportunity to harness Quantum AI's transformative power to revolutionize every aspect of our lives while safeguarding ethical standards and promoting a just society. It is both a challenge and an imperative—a call to action for global cooperation, innovation stewardship, and commitment to the greater good.

Balancing Innovation and Control

In the rapidly advancing frontier of quantum computing and artificial intelligence, the equilibrium between fostering innovation and maintaining control is paramount. This balance entails encouraging groundbreaking advancements while ensuring that complexities inherent in quantum AI do not spiral beyond regulatory and ethical bounds. As companies and research institutions race to harness the potential of quantum AI, policymakers must craft regulations that protect society without stifling the technological leap forward.

The heart of innovation lies in the unbridled exploration of possibilities. Quantum AI, with its promise of solving problems that classical systems struggle with, is on the brink of revolutionizing industries from healthcare to finance. However, blindly rushing towards innovation without any guardrails can lead to significant ethical challenges and unforeseen risks. Regulations serve as these guardrails, providing clarity and setting boundaries within which innovation can safely occur.

To craft effective policies, understanding the traits of quantum AI is essential. Quantum mechanics introduces a layer of unpredictability, emphasizing the need for adaptable and forward-thinking regulations. Traditional regulatory frameworks often fall short in addressing the nuances brought about by quantum phenomena such as superposition and entanglement. Thus, a hybrid approach that merges traditional regulatory practices with new, agile frameworks might be necessary.

Designing policies for quantum AI is a global undertaking. No nation or organization exists in isolation in this technological era. The interconnectedness of the world, especially in tech development, suggests a need for international cooperation in regulation. Global standards can prevent fragmentations and foster collaborative progress. They can also mitigate scenarios where leading nations in quantum AI impose their standards on others, promoting fair tech advancements worldwide.

However, global regulatory cooperation poses its own set of challenges. Diverse geopolitical landscapes and varying ethical values make it difficult to reach consensus. Countries have different priorities, with some emphasizing economic gain while others prioritize privacy and security. Nevertheless, the imperative for global dialogue and partnership remains critical for ensuring that quantum AI develops responsibly and uniformly worldwide.

In the domain of ethics and control, there's a pressing need for continuous dialogue between technologists and policymakers. Innovators must remain informed about regulatory requirements, while regulators should stay updated with technological advancements to avoid crafting outdated or irrelevant policies. Creating channels for collaboration helps bridge the knowledge gap, strengthens trust, and ensures that each party's concerns are addressed.

Public perception also plays a role in shaping policies. Policymakers must engage the public, whose trust in technology often hinges on how well it is managed and the perceived risks it carries. Initiatives for public education and awareness can demystify quantum AI, making people less susceptible to misinformation and more likely to support balanced regulations. Encouragingly, informed citizens can contribute valuable insights into the policymaking process, ensuring the public's interests are well-represented.

The balance between innovation and control also involves active monitoring and refinement of policies. Given the rapid iterations in quantum AI technology, regulations can easily become obsolete. Continuous assessment and iterative frameworks can help adapt regulations to reflect technology's evolving landscape. This approach, however, demands substantial resources and expertise, which may not be evenly distributed across nations or institutions.

In considering regulatory frameworks, one cannot overlook the crucial aspect of ethical AI. Biases in AI algorithms, questions of transparency, and the impacts on privacy and security are ethical concerns that take on new dimensions with the processing power of quantum systems. Striking a balance involves embedding ethical considerations into the development process of quantum AI from its infancy. Doing so ensures that AI systems are fair, accountable, and aligned with societal values.

Integrating ethics in policy also includes addressing the broader socioeconomic implications of quantum AI technologies. As these technologies alter the landscape of job markets, education, and industry standards, policies should guide equitable access and opportunities for all. Ensuring that quantum AI benefits society globally, rather than contributing to existing inequalities, calls for inclusive strategies and active engagement with diverse communities.

To conclude, balancing innovation and control in the realm of quantum AI is a complex yet vital endeavor. It requires collaborative international efforts, a constant dialogue between innovators and regulators, public engagement, and a steadfast commitment to ethics. By walking this tightrope carefully, we can unlock the immense potential of quantum AI while safeguarding humanity against its inadvertent challenges and disruptions.

International Cooperation and Standards

In the quantum AI domain, fostering international cooperation and establishing global standards isn't just an admirable goal—it's a necessity. The convergence of quantum computing and artificial intelligence holds great promise, but realizing its full potential depends on a connective tissue that spans continents and cultures. A cohesive framework enables nations not just to optimize technological development but also to maintain ethical standards and mitigate competitive disparities. The tapestry of international cooperation and standards provides a platform for countries to share insights, address potential challenges, and agree on common objectives.

Governments and organizations across the globe look at quantum AI with a dual lens of opportunity and responsibility. The technology's transnational nature demands that frontrunners like the

US, EU, China, and emerging tech hubs work in concert to navigate the intricate ethical and policy questions that arise. No single nation can address the complex challenges that quantum AI presents. This reality makes international collaboration a strategic imperative, where shared standards can ensure safety, interoperability, and equitable access to quantum-driven benefits.

Setting international standards for quantum AI is a Herculean task, but history attests to humanity's ability to collaborate on complex global issues such as aviation safety and nuclear energy. Drawing parallels to how foundational protocols were established in these fields, stakeholders in quantum AI can leverage similar frameworks to manage risks and facilitate innovation. The establishment of quantum-focused think tanks, regulatory bodies, and consortia can provide neutral forums where scientific discourse fuels policy formulation and standardization efforts.

But what do these standards look like? At a fundamental level, they should delineate guidelines around key aspects like data privacy, cybersecurity, ethical AI practices, and system interoperability. Compatibility is a particularly crucial area. Imagine a world where each country has its quantum AI stack with no regard for a universal protocol—chaos and inefficiency would ensue. Establishing a baseline for interoperability could involve creating open-source templates, standardized APIs, or even shared repositories of quantum algorithms that anyone, anywhere, can access and build upon.

One concrete example of international cooperation is the creation of multinational research hubs and innovation clusters. Initiatives like CERN for quantum computing could serve as centralized meccas for experimentation and development. These centers could be the breeding ground for quantum talents from across the globe, encouraging the cross-pollination of ideas and fostering a spirit of shared responsibility for technological evolution.

Inter-governmental collaborations, akin to the Paris Agreement for climate, could result in formalized treaties or accords stressing the ethical use of quantum AI. Such agreements would not only focus on the ethical but also prioritize environmental impact, given the considerable energy demands associated with quantum computing. Furthermore, these agreements could promote more affordable and equitable access to quantum resources, ensuring that lower-income nations aren't left behind in this technological renaissance.

The road to such cooperation isn't devoid of obstacles. Geopolitical tensions, disparate economic interests, and varying ethical frameworks can stall or complicate collaborative efforts. Countries are wary of losing competitive edges, which can hinder the free flow of information that is imperative for global standardization. Therefore, incentivizing participation is crucial. International bodies like the United Nations or the World Trade Organization could offer technology sharing agreements or sanction incentives to ease up cooperation among nations.

Moreover, bridging the gap between different biotechnical cultures remains a challenge, yet it is an opportunity. Quantum AI's impact will be broad, affecting sectors like healthcare, finance, and even agriculture. So, involving not only tech companies but also sector-specific stakeholders in the creation of standards will ensure that the solutions are as applicable as they are advanced. This interdisciplinary approach will align diverse views toward a cohesive mission—maximizing quantum AI's benefits while minimizing its risks on a global scale.

As these dialogues continue to unfold, ensuring that these standards are adaptable to rapid advancements in technology becomes pivotal. Technology's landscape shifts swiftly; regulatory frameworks and industry standards must be agile to keep up. An adaptable framework means not having rigid rules but rather flexible guidelines that encourage innovation while maintaining accountability. Such dynamism would allow nations to more easily revise policies to match the leaps that quantum AI makes.

In sum, international cooperation and standardization are the keystone elements in the responsible development and deployment of quantum AI. For tech-savvy individuals, professionals, and enthusiasts, understanding these multifaceted efforts is not just an academic exercise—it's a glimpse into the future governance of technologies that promise to reshape the way we live, work, and relate to one another. The stage is set, and global entities are fully equipped to act, ensuring that quantum AI becomes a tool for global enhancement rather than competition or disparity.

Chapter 22: Public Perception and Awareness

As we stand on the brink of an era defined by quantum computing and AI, understanding public perception and elevating awareness become paramount. Bridging the gap between intricate scientific concepts and the general public's understanding is essential for fostering an informed society. Educating a broader audience involves demystifying terms like superposition and neural networks, making them as relatable as everyday gadgets. While advances in technology promise transformative shifts across industries, misconceptions and unfounded fears often capture the spotlight. It's crucial to counter these with clear, evidence-based insights that emphasize the realistic potential of quantum AI. Integrating this knowledge into communal vernacular not only empowers individuals but also galvanizes collective enthusiasm for future possibilities. A well-informed public can support ethical deployment and regulatory frameworks with wisdom and foresight, ensuring that this technological revolution contributes positively to humanity's journey forward.

educating the General Audience

In an era marked by rapid technological advancements, understanding quantum computing and artificial intelligence (AI) is no longer a luxury reserved for specialists; it's becoming essential for the general public. The transition from highly specialized academic environments to mainstream discourse requires a targeted educational approach.

This means adopting strategies that simplify these complex topics without compromising their integrity, while also catering to various learning paces and styles.

When it comes to quantum computing, many people might feel intimidated by the alien nomenclature. Terms like "superposition" and "entanglement" can sound more like abstract philosophical constructs than components of a groundbreaking technology. But the key lies in relevance; if we can illustrate how these concepts directly influence daily life, the fog of complexity begins to lift. For instance, imagine explaining superposition with relatable metaphors like playing multiple musical notes at once or comparing entanglement to the emotional connections between twins, which remain strong even when they're apart.

AI presents its own challenges and opportunities for education. Much of the public already engages with AI technologies daily through digital assistants, recommendation systems, and smart devices, often without realizing the depth of underlying processes. The task is to translate technical constructs into everyday examples, demystifying AI by showing how it's an extension of human problem-solving capabilities. One could liken neural networks to human brain functions, helping people understand that AI learns in ways somewhat similar to how we do.

Utilizing multimedia educational platforms is crucial in making quantum computing and AI accessible. Interactive videos, gamified learning modules, and virtual reality experiences allow learners to explore these concepts in immersive environments. These tools not only present the material in a visually engaging manner but also encourage experiential learning—an approach particularly effective for subjects involving complex spatial and probabilistic elements, such as quantum circuits and AI decision trees.

Moreover, partnerships between educational institutions, tech companies, and public organizations can foster large-scale outreach programs. Initiatives that introduce quantum and AI concepts at school levels, for instance, can inspire future innovators and create a tech-aware society. Curricula must evolve to include fundamental lessons on these topics, giving students the edge in adapting to future job markets increasingly intertwined with advanced technologies.

However, education isn't just about disseminating information; it's about creating dialogue and curiosity. Public lectures, workshops, and seminars by leading figures in the fields can stimulate interest and encourage questions that drive forward collective understanding. Popular science books, blogs, and podcasts also serve as accessible resources, breaking barriers and inviting laypeople into discussions that shape future technological landscapes.

Another aspect of educating the general audience involves addressing misconceptions and fears surrounding these technologies. The idea that quantum computers might suddenly break all encryption or AI will overpower human roles often dominates public perception. Education can dispel myths by explaining where we stand in terms of current capabilities versus projected possibilities. By clarifying these points, we can replace fear with informed curiosity and critical thinking.

The democratization of knowledge regarding quantum computing and AI also touches on social and ethical dimensions. Who gets access to this education, and at what level? Efforts must ensure that learning opportunities are inclusive, reaching underserved communities that might otherwise be left behind by the tech evolution. Building such resources requires intentional policy and investment from stakeholders committed to equity.

History teaches us that public understanding of revolutionary technologies, like electricity or the internet, significantly impacts their development and adoption. As we stand at the brink of a new technological era, equipping the general audience with the knowledge to understand and evaluate quantum computing and AI becomes not just beneficial—it is necessary. The digital literacies of tomorrow depend on today's efforts to break down intimidating walls of jargon and present a world of possibilities in a language that's accessible to all.

By fostering a well-informed public, we can pave the way for collective decisions that steer quantum computing and AI toward enhancing human welfare globally. Empowered audiences can engage proactively with policymakers and innovators, promoting ethical standards and responsible implementation. These efforts can ensure that these groundbreaking technologies evolve as tools for the greater good.

Addressing Misconceptions

As the realms of quantum computing and artificial intelligence (AI) advance and intertwine, certain misconceptions have permeated public perception. It's not unusual for emerging technologies to be accompanied by misunderstandings, given their complexity and the revolutionary potential they promise. Addressing these misconceptions is vital to fostering informed discussions and encouraging responsible development and adoption of these cutting-edge technologies.

One of the most prevalent misconceptions about quantum computing is the notion that it will instantaneously render classical computing obsolete. While quantum computing does offer exponential speed-ups for specific problems, it's not a universal replacement. Classical computers excel at tasks that don't rely on quantum mechanical phenomena, and they will continue to do so in many applications. Quantum computers are set to complement classical computers, each addressing different types of computational challenges.

Similarly, there's often confusion about the capabilities of AI. Many believe AI can solve any problem or behave with human-like consciousness. In reality, AI systems are limited by their design and the data they are trained on. They lack the general intelligence and consciousness of humans. Understanding these boundaries is crucial to appreciating AI's current capabilities and potential.

The intersection of quantum computing and AI adds another layer of nuance. A common misconception is that quantum AI can autonomously solve complex problems without human intervention. Quantum algorithms offer powerful tools, but human insight and expertise remain essential in guiding their application. These technologies hold potential for symbiosis, not substitution.

There is also a widespread myth about quantum computers breaking all encryption overnight, leading to a catastrophic breach of security everywhere. While it's true that quantum computing could render some existing encryption methods obsolete, creating new quantum-safe encryption techniques is also underway. It is a race against time, and quantum cryptography stands as a promising shield.

Another misconception stems from science fiction and media portrayals: the fear of AI and quantum technologies leading to dystopian futures. Such narratives often exaggerate or misrepresent the technologies, casting them in a purely negative light. While addressing ethical and societal implications is essential, these technologies are tools that can be steered towards beneficial outcomes through careful governance and regulation.

The opacity of quantum mechanics often leads to misconceptions about what these technologies can and cannot do. Many envision a magical leap rather than a meticulous process of exploration and refinement that every scientific discipline undergoes. The reality of progress in quantum computing and AI is one of gradual advances, underpinned by rigorous testing, validation, and scaling challenges.

Addressing misconceptions also involves considering cultural and educational impacts. Misunderstandings may arise from gaps in education and awareness. It's important to integrate basic knowledge of quantum computing and AI into educational curriculums, enabling more individuals to make informed opinions and decisions concerning these technologies.

Moreover, understanding the economic and employment implications of quantum AI technologies is crucial. Fears of mass unemployment due to automation are partly rooted in misunderstanding the roles humans play in the innovation cycle. While some jobs may evolve or become obsolete, new roles and opportunities will emerge in tandem with technological advances.

The propagation of these misconceptions often owes much to sensationalist media coverage, which may emphasize speculative risks without adequately addressing the research community's efforts to mitigate them. Journalists and communicators have a responsibility to convey nuanced perspectives, offering not only potential risks but also realistic solutions and the incredible opportunities these technologies can present.

Ultimately, addressing these misconceptions is not about diminishing the awe-inspiring potential of quantum computing and AI but about providing a balanced and informed framework through which to understand it. Such a framework helps to inspire trust and motivation, encouraging collaboration across disciplines and industries to harness the full benefits these technologies offer.

As we nurture public perception and awareness, the role of educators, policymakers, and industry leaders becomes pivotal.

Ensuring transparent communication and fostering a culture of collaboration will build a more knowledgeable society poised to create and innovate responsibly.

In this transformative journey, raising awareness and dismantling misconceptions can pave the way for quantum and AI technologies to be tools for global good, ultimately solving challenges and enriching lives globally. Both caution and optimism are necessary as we venture into this exciting frontier, aiming to demystify and democratize the technologies that will shape our future.

Integrating Quantum AI Knowledge

The promise of quantum AI has sparked a new era of curiosity and innovation, captivating both the scientific community and the general public. The integration of quantum AI knowledge into the broader societal consciousness is a task that extends beyond technical training. It requires a holistic approach that considers existing perceptions, educational frameworks, and the outreach mechanisms that facilitate understanding.

Quantum computing, with its unique capacity to tackle problems previously deemed insurmountable, meets AI's revolutionary potential to transform the landscape of cognitive technologies. Yet, as with any transformative technology, the challenge lies not only in its development but also in how it is perceived and understood by non-specialists. Bridging this gap is crucial for fostering a more informed public discourse around quantum AI, thus aiding in its societal acceptance and eventual integration.

To raise public perception and awareness, it's essential to demystify the complexities surrounding quantum AI. Currently, due to its inherently technical nature, quantum AI is often shrouded in mystique. For many, discussions around qubits, superposition, or quantum entanglement might sound like the stuff of science fiction rather than scientific fact. Effective communication strategies should break down these concepts into accessible narratives, using analogies and examples from everyday life to make the abstract more tangible.

In education, incorporating quantum AI into curricula from as early as high school could lay the groundwork for a generation naturally fluent in these concepts. Just as computer science is now a staple of modern education, quantum AI should be presented not as an exotic branch of study, but as an integral part of technological literacy. By nurturing curiosity and encouraging exploration, educators and policymakers can equip students with the foundational skills necessary to engage with quantum AI more confidently and creatively.

Mass media and popular science outlets play a complementary role. Documentaries, articles, and podcasts that delve into the realms of quantum computing and AI are pivotal in reshaping perceptions. They serve as vehicles to disseminate knowledge beyond the academic and tech communities. Simplified explanations and compelling storytelling can spark widespread interest and inspire the next wave of thinkers and innovators.

Moreover, tutorials and interactive platforms can leverage the Internet's power to bring quantum AI knowledge into homes worldwide. Online courses offering free or affordable access combine flexibility and depth, reaching audiences who may not have the means to engage with formal academic institutions. Gamification of learning represents another frontier — educational games that simulate quantum processes can make learning interactive and fun, drawing in users through engagement and entertainment.

Public workshops and community engagement programs are another key mechanism for knowledge dissemination. Local and regional meetups hosted by universities or companies invested in quantum AI research create a grassroots approach to education. Here, enthusiasts, experts, and curious individuals can connect, share insights, and develop a communal understanding of what quantum AI can offer.

Amid these efforts, it's equally important to address misconceptions that might cloud public understanding. Science fiction often exaggerates or misrepresents the capabilities of quantum AI for dramatic effect, leading to both overblown expectations and unfounded fears. The portrayal of AI as either savior or villain is a narrative that needs balancing. Effective communication should focus on shedding light on the real, tangible benefits and potential risks of emerging technologies.

Transparency from developers and researchers will be vital in allaying fears and building trust. When the public can see and understand the motivations, challenges, and intentions of those at the cutting edge of quantum AI, it transforms skepticism into informed vigilance and cautious optimism. Open dialogues facilitated by conferences or public Q&A sessions can serve as forums where doubts and questions are addressed candidly, humanizing the technology and those behind it.

Ultimately, the endeavor to integrate quantum AI knowledge into public consciousness is a collaborative effort. Scientists, educators, communicators, and policymakers must work in tandem. Only through a concerted cross-sectoral approach can the potential societal benefits of quantum AI be fully realized. This collaboration not only nurtures public trust but also ensures that developments in this field align with societal needs and ethical considerations.

As we progress further into the quantum age, it becomes paramount to create an inclusive dialogue around these technologies. With the right educational tools and cultural narratives, we can prepare society at large to embrace quantum AI's advantages while understanding its limitations. This integration will not only enhance personal and professional growth but will also foster a workforce that's ready to harness the upcoming quantum revolution responsibly and innovatively.

The journey of integrating quantum AI knowledge is complex, yet it is one marked by unprecedented opportunity. We are poised at the precipice of a breakthrough where curiosity fuels understanding and understanding empowers action. In this landscape, informed public perception becomes a catalyst for innovation, ensuring that quantum AI's transformative potential reverberates through society, crafting a future where technology and humanity advance in harmony.

Chapter 23: Competing Technologies and Innovations

Amidst the surge of quantum computing and AI developments, a compelling narrative unfolds around the competing technologies and innovations that define this new era. While classical AI has driven transformative changes with its well-optimized algorithms and wide accessibility, quantum computing offers an alluring promise of groundbreaking enhancements, captivating the minds of technologists and futurists alike. The future prospects for traditional computing, boasting decades of evolutionary refinement, remain formidable, yet the fusion of hybrid solutions—blending classical and quantum paradigms—offers tantalizing benefits that can't be disregarded. These hybrid solutions are carving out niches where both can coexist, bringing extraordinary computational power and efficiency to bear on complex problems once deemed insurmountable. As this technological ecosystem evolves, a balance will emerge, steering innovations toward solutions that leverage the strengths of both worlds. The horizon is bright with promise, inviting exploration into uncharted territories where the line between vision and reality blurs, charting a course toward a future of unimagined potential.

Comparison with Classical AI

The landscape of artificial intelligence is ever-evolving, with quantum AI emerging as a potential game-changer. However, to understand its impact, we must first compare it to classical AI, which has been the foundation of our digital world. Traditional AI operates on classical computational principles, leveraging binary logic and sequential processing. This approach has led to remarkable advancements in machine learning, natural language processing, and neural networks. Yet, as powerful as classical AI has been, it faces limitations that quantum AI endeavors to overcome.

Classical AI algorithms are typically based on deterministic approaches. They rely on predefined pathways to find solutions, often requiring significant computational resources for tasks like image recognition or financial modeling. This deterministic nature means classical AI is constrained by the architecture of the classical computers they run on, primarily their binary state of zeroes and ones. As data grows exponentially and models become increasingly complex, these limitations become more apparent, necessitating innovative solutions.

A key distinction between classical and quantum AI lies in the fundamental principles of computation. While classical AI uses bits, quantum AI operates with qubits that leverage quantum superposition and entanglement. These quantum properties allow qubits to perform multiple calculations simultaneously, potentially offering exponential speedups for specific tasks. This parallelism means quantum AI can tackle problems that are computationally intractable for classical AI, transforming areas like cryptography, large-scale optimization, and complex system simulations.

The application of quantum algorithms to AI problems sets a new frontier. Consider Grover's algorithm, which provides quadratic speedups for unstructured search problems, or Shor's algorithm, which exponentially accelerates prime factorization. When applied to AI, these quantum algorithms can redefine what is computationally feasible, allowing for more sophisticated and nuanced analysis than classical approaches.

Despite these potential advantages, integrating quantum computing into AI isn't without its hurdles. Quantum systems currently suffer from issues such as error rates and coherence times, posing significant engineering challenges. Classical AI, in contrast, is mature, with robust frameworks and a vast ecosystem that spans industries. AI models like deep neural networks and decision trees are rapidly evolving, continuously improving thanks to a solid base of accumulated knowledge and resources.

Hybrid systems, combining classical and quantum methodologies, offer promising paths forward. These systems seek the best of both worlds, leveraging the strengths of classical AI while harnessing quantum enhancements where they are most effective. For example, classical pre-processing could be paired with quantum-accelerated solutions to manage large datasets efficiently, preserving the scalability of classical models with the added power of quantum computation.

Current classical AI systems also benefit from a familiarity that quantum AI lacks. Developers have spent decades honing AI algorithms and understanding their nuances. This expertise translates into practical applications that quantum AI, in its infancy, has yet to achieve on a broader scale. However, as researchers and industry pioneers push the boundaries of what's possible, quantum AI could revolutionize sectors that classical AI has only begun to touch.

Innovations in the classical AI realm are ongoing, with improvements in computational efficiency, the robustness of algorithms, and the incorporation of new techniques like meta-learning and generative adversarial networks. The advantage for classical AI is the existing infrastructure and rich legacy of research and application. These factors make classical AI a formidable benchmark for any emerging technology, including quantum computing.

Yet, the inspiration lies in what lies ahead. Quantum AI holds the promise of not just marginal improvements but rather a transformative shift, challenging existing paradigms by potentially reshaping complex systems analysis, enhancing pattern recognition capabilities, and solving optimization problems at unprecedented scales. This shift is why businesses and governments alike are investing heavily in quantum research, envisioning a future where quantum-enhanced AI could simulate molecular interactions, optimize logistics on a global scale, or revolutionize personalized medicine.

While classical AI has paved the way, quantum AI's evolution could lead to a new age of technological advancement. The competition and collaboration between these approaches underscore a vital part of technological progress: leveraging past achievements to build future innovations. As we continue to better understand quantum principles and improve quantum hardware stability and scalability, the integration of quantum computing into AI may extend the capabilities of human creativity beyond conventional boundaries.

In conclusion, the juxtaposition of classical and quantum AI isn't just a tale of competition; it's one of potential synergy. Classical AI has set a robust foundation, but quantum AI opens doors to possibilities that were once relegated to theoretical discussions. Bridging these two realms could lead to advancements that currently sit at the outermost fringes of our imaginations, positioning quantum AI as a key player in shaping a future defined by unprecedented technological possibilities.

Future Prospects of Traditional Computing

As the landscape of technology continues to shift with the advent of quantum computing and artificial intelligence (AI), the fate of traditional computing stands as a topic of notable intrigue. While some may hastily predict the obsolescence of conventional methods, those well-versed in the computing domain understand that such a transition is never straightforward. Traditional computing, anchored by classic von Neumann architecture, has been both the workhorse and the cornerstone of modern innovation for decades. Its future prospects, tied closely to ongoing advancements, adaptation, and integration into emerging technological frameworks, offer a resilient image that is often overlooked in futuristic discourse.

The adaptability of traditional computing platforms is one reason why they are expected to remain significant. Historical trends reveal their capacity to evolve, seen in the scaling of Moore's Law where computational power doubles approximately every two years. This trend, while facing physical and economic limitations, continues to inspire research into advanced processing technologies such as 3D integrated circuits and new semiconductor materials. These innovations present viable pathways for maintaining the relevance of traditional computing in the face of burgeoning quantum capabilities.

Moreover, traditional computing will not only coexist with quantum advancements but will also play a crucial supporting role. Quantum computers are not well-suited for every task; they're excellent at solving certain types of problems, thanks to their unique properties like superposition and entanglement, but not others. Traditional computing is expected to shoulder the everyday tasks where deterministic computation excels, such as running operating systems and managing applications that require predictable outcomes. This division of labor between classical and quantum systems underscores the complementary benefits of sustaining traditional computing approaches.

In many sectors, the integration of quantum computing technologies will themselves depend on traditional computing for interface and control operations. For instance, quantum computers will need high-performance classical systems to manage their notoriously complex inputs and outputs, offering a synergistic model rather than one of direct competition. The hybrid systems that combine classical and quantum computing components represent a nuanced evolution of technology that holds substantial promise.

Traditional computing also serves as a fertile ground for software innovation. Despite the emergence of quantum algorithms that promise speed-ups in certain complex computations, the majority of software development continues to occur on classical systems. Recent advances in artificial intelligence, facilitated largely by classical computing, are compelling examples of traditional computing's ongoing vitality. Tools such as machine learning models and neural networks, pivotal in today's AI research, rely heavily on the computational efficiencies provided by enhanced classical architectures.

Beyond pure technological perspectives, economic and ergonomic aspects also favor the continuation of traditional computing. The infrastructure around traditional computing—hardware production facilities, global supply chains, and professional expertise—represents a well-established ecosystem. Transitioning wholesale to quantum computing could demand resources and investments that dwarf current capabilities, suggesting that traditional computing offers a more immediate, financially viable solution for widespread application.

Educational pathways also contribute to the future prospects of traditional computing technologies. Since classical computing underpins much of current curricular frameworks, it lays the foundation for the next wave of innovators and technologists. As academia and industry continue to foster classical IT education, the knowledge and skills developed will play a critical role in both advancing traditional computing methodologies and assisting the integration of new quantum technologies.

Further, addressing societal needs and ethical considerations maintain traditional computing's relevance. It will continue to be integral in sectors requiring reliability and proven protocols, such as banking, healthcare, and governmental operations. These domains often demand stable and secure computing practices that traditional systems readily provide, ensuring their intrinsic value continues as new technologies unfold.

Ultimately, future prospects hinge on the ability of traditional computing to not just retain its utility but to redefine itself. As quantum and classical technologies converge, hybrid solutions seem increasingly viable. In these integrated systems, each component optimizes its strengths while carefully balancing its limitations. Traditional computing thus becomes an enduring partner in a technological landscape that's steadily diversifying with quantum breakthroughs at the forefront.

In conclusion, while quantum computing captures the imagination with its revolutionary potential, traditional computing remains indispensable. Its inherent strengths, adaptability, and the critical roles it continues to play across industries signify a future where classical and quantum systems coalesce rather than compete. This symbiosis promises an exciting era of exploration and growth that honors both the legacies of past innovations and the tantalizing possibilities of the future.

Hybrid Solutions and Their Benefits

As we stand on the brink of a technological renaissance, hybrid solutions emerge as a monumental bridge between two powerful realms: classical computing and the groundbreaking advancements of quantum computing. This fusion aims to harness the strengths of both worlds, leading to unprecedented problem-solving capabilities. While quantum computing alone offers immense potential in processing power and speed due to its unique characteristics like superposition and entanglement, it still faces practical challenges. Enter hybrid solutions—the perfect amalgamation that maximizes current technological infrastructure while integrating quantum's novel paradigms.

The core advantage of hybrid approaches lies in their ability to seamlessly integrate quantum capabilities with established classical computing techniques. By leveraging classical processors for tasks they perform efficiently and reserving quantum processors for specific, complex computations—such as optimization problems or simulations at scale—industries can enjoy both robustness and innovation. This marriage of technologies fosters greater efficiency, reducing limitations that either system would face independently.

For instance, in the burgeoning field of artificial intelligence, hybrid models can enhance machine learning algorithms. Classical AI techniques have achieved remarkable success in image recognition, language processing, and data analytics. Incorporating quantum algorithms into these models holds the promise of exponentially increasing the speed of learning processes, especially for tasks involving vast data sets or intricate pattern recognition. By synergizing the probabilistic nature of quantum computations with deterministic classical methods, AI can leap into uncharted territories of intelligence.

The benefits of hybrid solutions extend beyond speed and efficiency—they also offer a practical roadmap during the transition phase from classical to full-scale quantum systems. Companies burdened with legacy systems can incrementally integrate quantum components without a complete overhaul, which mitigates the risks associated with adopting a new, untested technology. This practical adoption path fosters innovation while maintaining operational continuity, an essential concern for industries wary of disrupting their business models.

Moreover, these solutions promise significant impacts on sectors like finance, where rapid data processing and intricate modeling are pivotal. Hybrid systems can perform complex calculations at a fraction of the time, refining risk assessments and optimizing portfolios with unparalleled precision. Such capabilities enable the anticipation of market changes with greater accuracy, empowering financial institutions to navigate volatile markets with newfound agility.

In the realm of cryptography, hybrid approaches leverage quantum mechanics to bolster security measures. Classical systems can handle standard encryption tasks, while quantum systems can perform secure key exchanges using quantum key distribution. This combination enhances the security framework, ensuring that sensitive information remains protected against potential breaches, especially as quantum threats loom.

Furthermore, hybrid computing models offer a viable strategy for environmental modeling and energy solutions. Classical computing can manage data collection and basic simulations, while quantum computing tackles large-scale, complex environmental models. This partnership could lead to advancements in climate change research by providing detailed simulations that help predict and mitigate environmental impacts more effectively.

In healthcare, hybrid solutions could revolutionize precision medicine by enabling the analysis of massive genomic datasets with greater speed and accuracy. Classical computational methods can manage patient data, while quantum processors can focus on intricate genetic data modeling, leading to more personalized treatment plans and faster drug discovery cycles.

This integration is not merely about technological compatibility; it's about creating synergy, extracting the best attributes from both paradigms. The flexibility and adaptive potential of hybrid systems cater to both short-term needs and long-term visions, ensuring that industries remain competitive and innovative.

Hybrid solutions also address one of the most pressing challenges in quantum computing—error correction. By using classical computers to oversee error checking and correction protocols, errors in quantum computations can be managed more effectively. This blending of technologies provides a cohesive framework that could extend quantum computing's range of practical applications.

Looking ahead, hybrid systems may be the stepping stone to a fully quantum future. They represent a strategic, measured approach to quantum implementation—a means to steadily break down existing barriers while inviting the future with open arms. The potential for hybrids to foster interdisciplinary collaboration is unprecedented, as they necessitate expertise from both the classical and quantum computing worlds, breaking down silos and encouraging a multidisciplinary approach.

This evolution paints a future that isn't just technologically advanced but is also accessible and practical for today. The constant innovation within these hybrid models points to a future where computational limits are not defined by the confines of hardware alone, but by the creative synergy of two technological titans working in harmony.

As we proceed into this new era of computation, hybrid solutions will likely serve as catalysts for ground-breaking discoveries and applications, transforming not just industries but the very fabric of our societal and economic structures. By embracing both classical and quantum strengths, we pave the way for a future where technology meets imagination, head-on.

Chapter 24: Long-term Vision for Quantum AI

As we look toward the horizon, the long-term vision for Quantum AI unveils an era where the boundaries between computation, intelligence, and human ambition blur into an unprecedented landscape of possibilities. Imagine a symbiotic relationship between machines and mankind, with quantum algorithms propelling artificial intelligence to tackle intricate global issues like climate change, disease eradication, and economic disparities. The coming decades might usher in a profound transformation in how societies operate, fostering environments where collaborative intelligence thrives, and decision-making reaches new heights. Yet, with this potential comes the challenge of navigating ethical dilemmas and technical hurdles. Roadmaps for future quantum AI development will need to prioritize equitable access, security, and the sustainability of advancing technologies. Ultimately, the enduring impact of Quantum AI will depend on our collective foresight and our willingness to harness its power for the greater good—transcending traditional limitations and unlocking insights that were once the realm of science fiction.

Predictions for the Next Decades

As we gaze into the horizon of technological advancement, the convergence of quantum computing and artificial intelligence presents a tapestry woven with both opportunity and transformation. The next few decades could see the dawn of an era where Quantum AI reshapes the very fabric of our daily experiences, industry practices, and societal structures. This unfolding landscape beckons with both promise and complexity, urging us to envision futures driven by machinery that defies the operational parameters of anything we've known before.

Quantum AI could radically accelerate problem-solving capabilities by leveraging principles of quantum mechanics such as superposition and entanglement. By breaking free from the binary limitations of classical computing, Quantum AI has the potential to perform complex calculations at speeds currently unimaginable. This shift could herald the minimization of computational time for solving multifaceted problems in fields like cryptography and optimization, thus unlocking solutions that have eluded human efforts for years.

In healthcare, we've long envisioned a future where medicine is deeply personalized. Quantum AI could be the key to achieving this vision, as it could analyze vast genomic data sets with unmatched speed and precision. The synthesis of AI's pattern recognition abilities with quantum computing's processing power might unveil new pathways for disease treatment and prevention. As such, Quantum AI could well be the cornerstone of bespoke medical treatments, tailoring healthcare to the genetic tapestry of individuals.

Taking a broader look, Quantum AI's automation capabilities could revolutionize industries across the board. Intelligent autonomous vehicles, trained with the ability to process immense amounts of environmental data instantaneously, could redefine transportation. Similarly, supply chains could operate with unprecedented efficiency, anticipating demand fluctuations and dynamically optimizing routes. Yet, while the gains in efficiency and effectiveness are clear, the societal impacts—particularly regarding employment—must be thoughtfully navigated.

Quantum AI's potential isn't limited to commercial and scientific sectors. In the realm of creativity, it could foster unprecedented collaboration between human and machine, giving rise to new forms of art unimagined by artists today. Music, visual arts, and narrative storytelling could see novel forms emerging that challenge our conventional understanding of creativity and authorship, reshaping the cultural landscape in ways both subtle and profound.

However, these predictions don't come without challenges. Quantum AI's future will grapple with issues of privacy and ethical use. As it becomes capable of processing and inferring from immense datasets, ensuring the protection of individual rights and maintaining ethical boundaries will be crucial. Institutions worldwide will need to craft policies that both harness the benefits of Quantum AI and mitigate its risks, fostering a balance between innovation and regulation that promotes public trust.

The educational sector must also evolve to keep pace with these technological advancements. Curricula that integrate quantum computing concepts will become essential, ensuring that upcoming generations understand and can operate within this new paradigm. Moreover, fostering interdisciplinary knowledge will be critical, as quantum computing affects a variety of fields. Hence, a cross-disciplinary approach to education will prepare future professionals to leverage Quantum AI's full potential.

In contemplating these vistas, we must also acknowledge the influence of geopolitical landscapes on the deployment of Quantum AI technologies. Nations are likely to vie for technological supremacy, much as they have for other major scientific advancements historically. International collaboration, however, will be vital in addressing global issues such as climate change, where Quantum AI's capacity for complex modeling could provide insights critical for sustainable solutions.

As we step forward, we find ourselves at the cusp of redefining intelligence itself. Quantum AI isn't just a leap in technological capability; it represents an evolution in how we conceive of problem-solving and creativity. The decades to come promise to test our adaptability and imagination, as we strive for a future where Quantum AI not only complements human endeavor but also transcends our current limitations, guiding us toward new frontiers that have yet to be charted. This journey is one of exploration, reflection, and innovation—a quest towards a future that whispers promises of possibilities waiting to be realized.

The Role of Quantum AI in Future Societies

As we stand on the brink of a technological revolution, the convergence of quantum computing and artificial intelligence (AI) is poised to fundamentally reshape the societies of tomorrow. Quantum AI, an emergent field at the intersection of these two powerful domains, holds the potential to address complex challenges across various facets of human life. The transformative power of Quantum AI stems not just from enhanced computational capabilities but also from a paradigm shift in how we perceive and tackle problems. This new approach could redefine the limits of what's achievable, enabling breakthroughs that were once the realm of science fiction.

One of the most compelling roles of Quantum AI lies in its potential to drive unprecedented advancements in healthcare. Quantum-enhanced algorithms promise to decode the vast and intricate data sets associated with human biology, paving the way for insights into diseases that have long eluded researchers. Potential personalized treatments, tailored to the genetic makeup of each individual, could revolutionize medicine, turning once-deadly conditions into manageable ones. This technology could bring about a world where health systems not only treat but predict and prevent illness, fundamentally altering our relationship with wellness and longevity.

Beyond healthcare, Quantum AI could play a pivotal role in addressing some of the most daunting environmental challenges of our age. Climate change, resource scarcity, and biodiversity loss are complexities that demand innovative solutions. Quantum AI could offer sophisticated climate models that surpass current predictive abilities, enabling more accurate forecasting of climate phenomena and modeling of sustainable energy systems. By optimizing everything from energy networks to resource distribution systems, Quantum AI could contribute significantly towards building a sustainable future.

In the realm of finance, Quantum AI's impact may be equally groundbreaking. It could enhance risk assessment, optimize investment portfolios, and detect fraudulent activities with unprecedented precision. This could lead to a more stable and efficient financial system, balancing risk and return better than ever before. Financial institutions might harness these capabilities to redefine banking, ensuring security and transparency in every transaction.

Quantum AI's influence extends further into the sphere of transportation. It could revolutionize logistics and traffic systems through the application of complex optimization models that surpass classical capabilities. Autonomous vehicles could become safer and more efficient, navigating real-time traffic patterns and minimizing accidents significantly. It's not just about making transportation faster and more efficient; it's about making it smarter, responsive to the real world in ways never before possible.

At the societal level, Quantum AI could redefine the nature of work and professional life. With advanced automation and augmentation of human capabilities, Quantum AI might shift the focus from traditional repetitive tasks to creative and strategic endeavors. This evolution has the potential to spark transformations across numerous industries, fostering innovation and efficiency. However, it's also vital to address the societal disparities that could emerge from such shifts, ensuring that the benefits of Quantum AI are equitably distributed.

Education stands as another key area where Quantum AI could dramatically alter the landscape. By enabling personalized learning experiences, it could ensure that educational content is tailored to the unique needs of each student, expanding opportunities for learning in unprecedented ways. Quantum AI could support educators by providing valuable insights into student progress and tailored recommendations, fostering an environment conducive to lifelong learning and adaptability.

As Quantum AI technology continues to evolve, ethical considerations must remain at the forefront of its development and integration into society. Responsible AI development involves addressing potential biases, ensuring transparency, and creating regulatory frameworks that balance innovation with control. Global cooperation could be crucial in establishing standards and policies that govern the use of Quantum AI, promoting its potential for good while mitigating risks.

The role of Quantum AI in future societies is far-reaching and multifaceted. Its impact isn't confined to a single domain but touches on various aspects, from healthcare and finance to transportation and education. As these technologies mature, they promise to unlock possibilities we haven't yet imagined, driving us towards a future where the boundaries of what's possible are constantly expanding. Yet, the journey towards this future demands careful navigation, with considerations of ethical, social, and technological dimensions guiding the way forward.

Challenges Ahead and Roadmaps

In weaving a long-term vision for Quantum AI, one can't overlook the myriad challenges that lie ahead. As we peer into this bold and transformative future, it's crucial to identify the obstacles that could slow or even derail progress. From the technical intricacies and ethical quandaries to the economic implications, these challenges are as diverse as they are complex. What's exciting, however, is that each challenge presents an opportunity for innovation, urging us toward unprecedented advances that might redefine human potential.

First, the technical challenges dominate the landscape. The quantum hardware required for sophisticated AI tasks is still in its infancy, and scalability remains a significant hurdle. Current quantum computers are not yet robust enough for practical, large-scale AI applications. Building and maintaining extensive qubit systems, ensuring coherence, and mitigating error rates demand extensive research and resource allocation. Addressing these issues involves groundbreaking developments in qubit stability and error correction techniques, so researchers must stay persistent and innovative in their approaches.

Beyond technical barriers, ethical considerations present another monumental challenge. With Quantum AI's potential to process and analyze personal data at unprecedented speeds and accuracy, ensuring privacy and preventing misuse is paramount. The confluence of quantum computing and AI raises concerns of surveillance, data exploitation, and decision-making transparency. Therefore, a crucial roadmap involves setting ethical guidelines and regulatory frameworks keeping pace with technological advancement, ensuring they do not trample upon basic human rights.

Economic challenges too, loom large on the horizon. Solving complex computational problems using Quantum AI isn't just a technical and ethical venture but a financial one. The costs of developing, deploying, and scaling Quantum AI technology could be prohibitive. This raises questions about accessibility and adoption by different sectors and nations. A pragmatic roadmap requires figuring out how to democratize Quantum AI, enabling equitable distribution of its benefits across the socio-economic spectrum. Collaborations, public-private partnerships, and government incentives could be vital components of this journey.

Another intriguing aspect to consider is the skill gap. Deploying Quantum AI on a large scale will necessitate an interdisciplinary workforce equipped with unique skills. As it stands, there aren't enough experts well-versed in both quantum mechanics and advanced AI. This skill mismatch could create bottlenecks hampering widespread adoption. It's imperative to invest in educational programs that cultivate a new generation of technologists adept in these dual domains. Overcoming this demands a robust educational roadmap to align academic curricula with industry needs, promoting lifelong learning opportunities and encouraging diversity in STEM fields.

Innovation ecosystems and collaborative research efforts must be prioritized as formidable strategies within this roadmap. Sharing discoveries and developmental insights through international cooperation can transcend individual limitations, propelling collective progress. Global partnerships play an essential role in creating research hubs that are talent rich and resource abundant, nurturing an environment where ideas cross-pollinate to yield groundbreaking outcomes.

Emerging on the horizon are possibilities of unexpected breakthroughs and applications. Yet, weaving these breakthroughs into our societal fabric is another challenge altogether. While Quantum AI could lead to advancements in healthcare, environmental sustainability, and beyond, the transition requires careful thought. Integrating these technologies means rethinking existing infrastructures and societal norms, which often resist rapid changes. Crafting roadmaps in this area might involve phased adoption strategies, ensuring communities understand and welcome these innovations.

Lastly, it's paramount to address the psychological barriers and public perception challenges tied to such disruptive technologies. Public distrust or misunderstanding can stifle growth and innovation efforts. Building awareness and fostering a narrative that demystifies Quantum AI, underlining its benefits and addressing its concerns, should be well-articulated within the roadmap. This involves communications strategies that are transparent, independent validations from trusted sources, and educational outreach programs that speak to audiences both technical and general.

Indeed, as we envision a future powered by Quantum AI, it's noted that challenges are not mere roadblocks but avenues that demand ingenuity, resilience, and collaboration. Roadmaps unfolding in this landscape must be dynamic, prepared to accommodate unforeseen discoveries and disruptions. In confronting these challenges, the Quantum AI community has the potential to navigate the unknown with a blend of caution and boldness, ensuring that these technologies contribute positively to the global community.

While the path of Quantum AI is fraught with trials, it is also paved with potential. Designing and following a structured, comprehensive roadmap will be essential to transforming future visions into present realities. It's the meticulous mapping of routes through this intricate terrain that holds the promise of a new era, where what seems like the realm of fiction today becomes the bedrock of our society tomorrow.

Chapter 25: Collaboration Across Disciplines

The transformative potential of quantum computing and AI is at its peak when these technologies meld with various disciplines, sparking unprecedented innovation. In this interconnected era, collaboration isn't just beneficial; it's imperative. Scientists, engineers, ethicists, and policymakers increasingly find themselves in the same room, hashing out shared goals that were once thought disparate. This interdisciplinarity fosters a fertile ground where quantum algorithms meet AI-driven insights, leading to novel solutions that address complex global issues—from climate change to healthcare innovation. Moreover, by weaving together diverse threads of expertise, we expedite the development of robust frameworks and scalable solutions that were previously unattainable. As lines blur between fields, a new generation of technologists is emerging, equipped with a holistic understanding and capable of navigating these intricate landscapes. This dynamic confluence paves the way for a future where, powered by synergy, the frontiers of possibility are collectively pushed forward. Collaboration across disciplines is more than wise—it's our strongest catalyst for revolutionary progress.

Interdisciplinary Research Initiatives

In an age where technological boundaries continually expand, the fusion of disciplines has become more than just a trend; it's a necessity. Interdisciplinary research initiatives tap into the core of innovation, especially when it comes to pioneering technologies like quantum computing and AI. No longer can one discipline, no matter how advanced, claim to hold all the keys that unlock the vast potential these technologies offer. Instead, it is through the intricate dance of integrating diverse fields that we begin to unveil truly revolutionary advancements.

At the forefront of this interdisciplinary exploration is the marriage between quantum physics and computer science. The challenges in quantum computing demand a robust understanding of not just physics but also algorithm design and computational theory—fields traditionally seen as separate or complementary. However, when these fields intertwine, they create novel approaches to solve complex problems, surpassing the capabilities of classical computers. This synergy is the catalyst for quantum algorithms capable of exponential speedups, solving problems intractable with current technology.

It's not just physics and computer science where boundaries blur and spark innovation. Consider the intersection of neuroscience and AI. Here, interdisciplinary research is pivotal for artificial intelligence systems that aim to model and mimic human cognitive functions. Neuroscientists, psychologists, and computer scientists collaborate to decode the human brain's mysteries, applying these insights to develop machine learning models and neural networks that operate with enhanced efficiency and accuracy.

Moreover, the delicate and complex interactions between different scientific and engineering disciplines create a robust landscape for innovation. In materials science, chemists and physicists work alongside computer scientists to devise quantum materials that may better support quantum entanglement and superposition. Similarly, interdisciplinary collaboration in fields like mathematics and computational biology is opening new frontiers in understanding biological systems through the lens of quantum mechanics.

Quantum-based drug discovery stands as an exemplary field demonstrating the power of interdisciplinary initiatives. By combining quantum computing with molecular biology, researchers are developing simulations that predict molecular interactions in a fraction of the time compared to traditional methods. This approach can accelerate the development of new, more effective medications, with impactful implications for disease treatment and healthcare.

In this interconnected research environment, we can't overlook the role of social sciences in guiding technological growth. Interdisciplinary research that includes ethics, philosophy, and sociology is vital to ensure responsible development processes in AI and quantum technologies. These human-centered perspectives drive the discussion on the ethical use of technology, emphasizing the importance of considering societal impacts alongside technical advancements.

Interdisciplinary research initiatives also find a crucial partner in educational institutions. Universities and research organizations are rethinking curricula to break down traditional disciplinary silos. They've started offering programs that blend quantum computing with AI, fostering new generations of scientists and engineers who think cross-disciplinarily. These forward-thinking educational approaches are producing a workforce ready to tackle tomorrow's daunting challenges, equipping them with the holistic knowledge and skills necessary for interdisciplinary innovation.

The role of collaborative platforms and funded research networks in fostering interdisciplinary initiatives cannot be understated. International research collaborations and consortia exemplify how pooling expertise from various domains accelerates innovation. Through shared knowledge and resources, institutions can undertake large-scale projects that would be impossible individually. This spirit of cooperation is essential for translating complex theories into tangible technological breakthroughs.

Government and industry partnerships also feed into this dynamic, providing resources and platforms necessary for sustaining interdisciplinary research. Industry brings practical insights and funding, while academic research supplies theoretical depth. Together, they create ecosystems where research can swiftly transition from hypothetical models to real-world applications.

The path forward in interdisciplinary research initiatives is not without challenges. Aligning methodologies, managing diverse teams, and harmonizing intellectual contributions require deliberate efforts and leadership. Communication barriers, differing terminologies, and disciplinary biases frequently threaten these collaborations. However, with the right frameworks and a culture of openness, these hurdles are surmountable, leading to an enriching and productive confluence of ideas.

In conclusion, interdisciplinary research initiatives hold the promise to unravel the complexities associated with quantum computing and AI, pushing beyond the frontiers of what one discipline can achieve alone. By merging talents, knowledge bases, and intellectual traditions from a mosaic of fields, we set the stage for a future rich with innovation and discovery. In this dance of collaboration, we not only strengthen our ability to solve entrenched scientific problems but also better equip ourselves to harness the transformative power of technology responsibly and ethically.

Combining Expertise for Greater Impact

In the rapidly evolving landscape of technology, combining expertise across different disciplines is not just advantageous—it's essential. The intersection of quantum computing and artificial intelligence (AI) serves as a vivid testament to the power of collaborative innovation. By harnessing the strengths of multiple fields, we can propel technological advancements to new heights, addressing complex problems and unlocking unprecedented opportunities.

Consider the integration of quantum physicists, computer scientists, AI experts, and domain-specific specialists working together toward a common goal. This kind of synergy accelerates progress in ways that isolated efforts rarely can. Quantum computing, with its ability to process information at speeds unimaginable with classical computers, offers a fertile ground for AI applications that require vast computational resources. This collaboration not only amplifies the capabilities of AI systems but also reinforces the foundational principles of quantum mechanics.

The melding of these disciplines is akin to a carefully orchestrated symphony, where each instrument contributes to a harmonious whole. AI, for instance, can be instrumental in controlling quantum systems, which often exhibit behaviors that are difficult to interpret with traditional methods. Through machine learning, AI can predict system behaviors, optimize quantum algorithms, and even experiment with new quantum states, further enhancing the field's knowledge base.

One primary driving force behind interdisciplinary collaboration is the recognition that no single discipline has a monopoly on innovation. By bridging the gap between physics and computer science, with an infusion of insights from various application domains like healthcare, finance, and environmental science, the possibilities become boundless. Such collaborations mean that each discipline can contribute its unique perspective, leading to more holistic solutions to pressing challenges.

Moreover, the integration of diverse expertise contributes to the development of cutting-edge tools and methodologies. For instance, quantum-enhanced neural networks represent a tangible outcome of the collaboration between quantum computing and AI. These networks are designed to utilize quantum properties to enhance learning processes and computational efficiency, potentially revolutionizing how data is analyzed and interpreted across various industries.

As we delve deeper into the potential of quantum AI, it's crucial to recognize that the greatest breakthroughs often lie at the fringes of what is currently understood. By cultivating interdisciplinary teams, organizations can push these boundaries, venturing into new territories with the confidence that comes from a collective pool of knowledge and experience. The concept is simple: the sum of these combined efforts is far greater than its individual parts.

Further, combining expertise across disciplines encourages a culture of continuous learning and adaptation. In a world where technology is evolving at breakneck speed, staying at the forefront requires a commitment to learning from others and integrating new perspectives. This habit of learning not only spurs innovation but also ensures that the solutions developed are robust, sustainable, and relevant to the real-world problems we aim to solve.

In practical terms, fostering such collaborations demands an environment where communication is fluid, and barriers to cross-disciplinary interaction are minimized. This means creating platforms for dialogue and incubation spaces where professionals from different fields can collaborate hands-on, exchange ideas, and inspire one another. It is equally important to build networks that connect experts globally, transcending geographical constraints and fostering a global community of innovators.

As we look ahead, the importance of collaboration across disciplines cannot be understated. With a multidisciplinary approach, the very fabric of technological advancement is enriched with a diversity of thoughts and methods. The fusion of different academic and practical perspectives catalyzes innovative thinking, driving major leaps in understanding and capability.

Ultimately, the collaboration across disciplines isn't just about achieving greater technical results; it's about shaping a future where technology can be used as a force for good, addressing complex societal needs with precision and empathy. As we continue to explore the interplay between quantum computing and AI, investing in multidisciplinary efforts will be key to realizing their full potential for greater impact, shaping a future where technology serves humanity in the truest sense.

Building a Multidisciplinary Workforce

The future of technology is reshaping with the synergy between quantum computing and AI, and at the heart of this transformation lies the necessity for a multidisciplinary workforce. The confluence of distinct yet complementary fields like physics, computer science, mathematics, and cognitive science calls for professionals who can navigate this diverse landscape. Building such a workforce requires not only a strategic approach to education and training but also an environment that fosters collaboration across different domains.

A fundamental aspect of cultivating a multidisciplinary workforce is the integration of diverse skill sets. In the realm of quantum computing, an understanding of quantum mechanics is just as crucial as proficiency in algorithm development. Similarly, developing AI solutions that leverage quantum computing requires insights from both AI specialists and quantum physicists. Therefore, educational institutions and organizations must emphasize cross-disciplinary learning, encouraging students and professionals to acquire skills outside their core area of study.

One approach to addressing this need is the introduction of interdisciplinary programs that incorporate elements of both quantum computing and AI. Such programs should feature courses that cover the essentials of each field while highlighting their intersection points. For instance, a course might explore how quantum algorithms can enhance machine learning models, providing students a platform to experiment with innovative techniques. This hands-on approach can spark creativity and foster a deeper understanding of the complex, intertwined nature of these technologies.

Apart from formal education, fostering a multidisciplinary workforce also involves creating opportunities for collaboration. Research labs, technology hubs, and innovation centers should serve as melting pots where professionals from varied disciplines can exchange ideas and work on joint projects. This culture of collaboration can drive innovation as different perspectives often lead to revolutionary breakthroughs. The interaction among experts can lead to a collective problem-solving approach where challenges are tackled from multiple angles simultaneously.

Moreover, industries can benefit significantly from hiring teams that embody diversity in thought, experience, and education. A diverse team synthesizes different problem-solving methods, leading to more comprehensive and creative solutions. For technology companies focusing on quantum AI, employing individuals with backgrounds in fields such as cognitive psychology or material science may introduce novel insights and potential applications. Employers should therefore aim to create inclusive work environments that digress from conventional hiring practices and look for talents with unconventional experiences and backgrounds.

It's crucial to recognize that the role of soft skills in a multidisciplinary workforce is equally important. Communication, collaboration, and adaptability are key attributes that enable professionals to work productively across strict academic or technical boundaries. Training programs should thus incorporate modules focusing on enhancing these skills. Effective communication ensures that complex ideas can be translated and shared across departments, while adaptability allows professionals to transition between roles and projects that may require different skills and knowledge bases.

Additionally, mentorship and continued professional development are vital in nurturing a resilient multidisciplinary workforce. By providing access to mentors who possess cross-disciplinary expertise, organizations can guide employees through their career progression while simultaneously empowering them to pioneer innovation. Continuous learning initiatives will help professionals stay current with advancements in their respective fields as well as in adjacent ones.

The role of policy and management in establishing this diverse workforce cannot be overstated. Organizations must design policies that support continuous learning and cross-functional training. Such policies could incentivize employees to seek further education or participate in external projects that enhance their multidisciplinary skills. Moreover, management practices that encourage risk-taking and non-traditional thinking can help unlock the potential of a multidisciplinary workforce.

Furthermore, the shift towards a multidisciplinary approach needs to happen at both macro and micro levels. Governments and educational bodies should promote curricula that blend subjects which traditionally stand alone. At the micro level, companies can lead by example, integrating multidisciplinary methodologies into their project design and problem-solving frameworks.

International collaboration, too, stands as a cornerstone for building an effective multidisciplinary workforce. Quantum computing and AI are global endeavors, and fostering relationships across borders can allow for knowledge exchange and collaborative research on an unprecedented scale. Policies that support international education and professional exchange programs can bridge gaps and build networks that solve shared challenges in technology and science.

In sum, building a multidisciplinary workforce is not a mere trend; it's a necessity as we venture further into the age of quantum AI. It requires strategic planning and commitment to education, policy-making, and creating collaborative environments. With the right investments in this direction, we can inspire a generation of professionals equipped to tackle the complex, multifaceted challenges of tomorrow.

Conclusion

As we reach the conclusion of this comprehensive exploration into the realms of quantum computing and artificial intelligence, it's clear that we're standing at the precipice of a technological revolution. These two fields, individually monumental in their impact, are increasingly intertwined, hinting at a future where their combined potential reshapes the very fabric of our lives. The journey through the various dimensions of quantum AI has been a testament to the relentless pursuit of knowledge, innovation, and the desire to harness these advanced technologies for the betterment of society.

Throughout this book, we delved into the intricacies of quantum mechanics, a field that challenges our classical understanding of computation and offers opportunities once deemed science fiction. Quantum bits, or qubits, have given us a window into computing processes that can operate with unprecedented speed and efficiency. In parallel, we explored the foundations and evolution of artificial intelligence, a domain where machines learn, adapt, and execute tasks mimicking human intelligence. These technologies are not just the culmination of complex algorithms and physics but also a testament to human ingenuity and curiosity.

At the intersection of quantum computing and AI, transformative potential awaits. Quantum algorithms are poised to solve intricate problems much faster than classical systems can manage. This isn't merely an academic exercise; it's a leap toward answering questions in fields as diverse as healthcare, finance, transportation, and beyond.

From optimizing traffic systems with precision to revolutionizing drug discovery processes, quantum AI promises a future that augments human capability and expands our horizons.

However, with great power comes great responsibility. The ethical considerations surrounding quantum AI are profound and multifaceted. As these technologies develop, ensuring they are deployed responsibly is paramount. Addressing biases within AI systems, creating robust regulatory frameworks, and fostering international cooperation will be critical steps in mitigating potential negative impacts. Responsible innovation will require vigilance and collaboration across governments, industries, and academia.

The impact of quantum AI on the job market is another area ripe for discussion. While some roles may become obsolete, new opportunities will arise, necessitating a workforce that is adaptable and willing to embrace change. Education and continuous learning will be vital in preparing individuals for jobs in a quantum AI-driven world. By nurturing interdisciplinary skills, we can help shape a future where humans and machines coexist and collaborate fruitfully.

Looking ahead, the predictions for quantum AI are both exciting and daunting. As quantum systems become more scalable and error correction techniques improve, the practical applications of these technologies will expand exponentially. The societal implications will be vast, influencing everything from how we interact with technology to how we perceive intelligence itself. This evolution requires forward-thinking policies that balance innovation with ethical considerations.

Ultimately, the long-term vision for quantum AI involves a profound transformation of our technological landscape. With ever-increasing global collaborations, these advancements will foster shared prosperity and encourage solutions to some of the world's most pressing problems. The road ahead is not without its challenges, from technical hurdles to societal adaptation, yet the potential rewards are immense. As we move forward, embracing change and fostering an environment of creativity and innovation will be central to realizing this vision.

In summary, the path of quantum computing and AI intertwines with nearly every aspect of modern society. Our exploration has only begun to scratch the surface of what's possible. The future, rich with possibility, calls for continued exploration, robust dialogue, and inclusive decision-making. With informed optimism, we can step forward into an era where quantum AI not only enhances the human experience but also redefines what it means to be intelligent in a quantum world.

Appendix A: Appendix

In this appendix, we aim to provide supplementary information and resources that complement the primary content of the book, enhancing the reader's understanding and engagement with the topics of quantum computing and AI. While the main chapters offer comprehensive insights into the intersection of quantum technology and artificial intelligence, this section serves as a repository for additional materials and summarizes key resources to further support your exploration and learning journey.

Glossary of Terms

To aid in the comprehension of technical jargon used throughout the book, a glossary of terms is provided here. These definitions are intended to be concise yet thorough, ensuring clarity as you navigate complex concepts related to both quantum computing and AI.

Qubit: The fundamental unit of quantum information, analogous to a classical bit but capable of existing in multiple states simultaneously.

Superposition: A quantum phenomenon where a qubit can exist in multiple states at once, allowing for complex computations.

Entanglement: A unique quantum property where pairs of particles become linked, and the state of one instantaneously influences the state of the other, regardless of distance.

Further Reading and Resources

For readers eager to delve deeper into specific aspects of quantum computing and AI, we have curated a list of books, articles, and online courses. This section provides pathways to broaden your knowledge and engage with current research trends.

Quantum Computing for Computer Scientists by Yanofsky and Mannucci: An essential resource for understanding the computational framework and capabilities of quantum computing.

Artificial Intelligence: A Guide to Intelligent Systems by Michael Negnevitsky: A comprehensive introduction to the principles and applications of AI technologies.

Online Courses: Platforms like Coursera and edX offer specialized courses developed by leading universities, providing structured learning paths tailored to both beginners and advanced learners.

Common Questions and Misconceptions

As innovative as quantum computing and AI are, they often come with questions and misconceptions. This subsection addresses common inquiries and offers clear explanations to correct misunderstandings that might arise from the rapid advances and occasionally sensationalized media reports.

Can quantum AI solve any problem instantly? While quantum-enhanced AI has the potential to tackle certain problems much faster than classical AI, it is not a panacea for all computational challenges. Practical limitations and ongoing research continue to define its capabilities.

Is AI going to replace all human jobs? AI's role is primarily to augment or enhance human capabilities rather than replace them completely. While some jobs may become obsolete, new opportunities and roles are likely to emerge, requiring a shift in skills and workforce adaptation.

Contact and Community Engagement

Engagement with the community is vital for shared growth and learning. We encourage readers to participate in forums, attend conferences, and collaborate in research initiatives related to quantum computing and AI. Several online communities provide platforms for discussions, idea exchanges, and networking:

Quantum Computing Stack Exchange: A question-and-answer site for practitioners and enthusiasts.

AI Alignment Forum: A place for in-depth discussions about AI development, ethics, and future strategies.

We invite readers to connect with the authors via institutional or professional channels to discuss the content, explore opportunities for collaboration, or share insights and feedback. Together, we can continue to navigate the evolving landscape of technology and build foundations for a future enriched by the intersection of quantum computing and AI.

www.ingramcontent.com/pod-product-compliance
Lightning Source LLC
Chambersburg PA
CBHW051227050326
40689CB00007B/828
* 9 7 8 1 4 5 6 6 5 8 7 9 3 *